I am interested to see how Nature proceeds to heal the wounds.

Henry David Thoreau, *Journal*, 1855

HEALING NATURES, REPAIRING RELATIONSHIPS

New Perspectives on Restoring Ecological Spaces and Consciousness

Edited by
Robert L. France

Library of Congress Control Number 2007922452
ISBN 0-9717468-6-9
ISBN13 978-0-9717468-6-2

First Edition

Green Frigate Books
P.O. Box 461
Sheffield, Vermont
USA 05866-0461
www.greenfrigatebooks.com
info@greenfrigatebooks.com

France, Robert L. (Editor)
Healing Natures, Repairing Relationships: New Perspectives on Restoring Ecological Spaces and Consciousness.
LCCN: 2007922452
ISBN: 0-9717468-6-9
ISBN13: 978-0-9717468-6-2

1. Environment. 2. Ecological Restoration. 3. Nature Philosophy. 4. Landscape Architecture

Front cover: Boston's Back Bay Fens, one of the first restoration design projects created at the end of the nineteenth century for wetland restoration, sanitation treatment and public recreation.

Back cover: Children enjoying a contemporary restoration design project near Vancouver created for wetland restoration, stormwater management and public recreation.

Printing this book on recycled stock has saved:

13.1 trees (40' tall and 6-8" diameter); 5,554 gallons of water; 2,234 kilowatt hours of electricity, 612 pounds of solid waste; 1,203 pounds of greenhouse gases.

CONTENTS

Prologue

Swamped! A Tale of Two Restorations: Part I: The View From Home

Robert France

"Redeeming a swamp....comes pretty near to making a world"

—Henry David Thoreau, 1854, *Journal, in* France, R. L. *Wetland Design: Principles and Practices for Landscape Architects and Land-Use Planners,* 2003

Putting down the collection of international newspaper clippings about restoration, I stare out my study window at the developed landscape of the neighborhood. The scene is one of residential domestication, nature experienced only occasionally in the form of birds flying by overhead on their way to someplace else or suburban "pests" rummaging around in backyard containers of discarded waste. If one could see through the eyes of one of those birds and look down on north Cambridge, Massachusetts, the overall impression of the area would be of a landscape covered by a sprawling sea of gray pavement, roofs and other hard surfaces.

It was not always this way of course; for the area had once been the center of the "Great Swamp," one of America's most historic wetlands. For hundreds of years, native Amerindians set up their nets to catch the vast numbers of alewife and other fishes moving up river from the coast and into the expansive tidal wetlands during spring spawning runs. These marsh dwellers would also use the few elevated rises as staging areas for hunting the abundant animal life that then existed in the wetland. The wetland also provided material for construction of their homes and boats. Years later, Thoreau would pause in the area for his "botanizing" diversions on trips between Boston and his home in nearby Concord. It was here that academic ornithology had its roots in North America through the natural history investigations of Harvard University, and some of the very last passenger pigeons were seen in the remains of the wetland.

"Remains," because the Great Swamp is in a sense a microcosm for every deleterious action that humans have ever inflicted upon wetlands: ditching, diking, dredging, damming, draining, and finally destroying. Endemic culture was also lost during this process, for, in a scene reminiscent of the "purchase" of Manhattan, it was on the shores of this wetland where the transaction between the colonists and First Nation peoples took place with the signing over of vast tracks of the land (essentially much of present-day Cambridge and neighboring cities) to the Europeans. Following the progressive and relentless expansion of early colonial farms into the area (some of the first inland European settlements in the country), the birth of one of the first American ventures in international business (ice harvesting from the kettle lakes that fringe the wetland) brought numerous rail lines to the area. Then, in the middle of the nineteenth century, famine-fleeing Irish, prevented from settling in more respectable neighborhoods in Boston, were relegated to the city fringes where they worked in the brick-making business, exploiting the thousands of years of deposited alluvial clays. The connection of a tidal stream to one of the ice ponds was severed and the stream interred underground. No matter, since a dam was soon constructed downstream that for the first time removed all influence of tidal water to the wetland as well as ending the seasonal migration of anadromous fishes. Malarial fears at the dawn of the twentieth century led to placing the hydrologically effective sinuous rivers into hydraulically efficient concrete channels. Situated between the borders of four small cities operating under the political "home rule" doctrine of the Commonwealth of Massachusetts, the area began to accommodate all the industries and activities deemed inappropriate for the central community hubs. Starting with abattoirs and tanneries, then steel yards, and then chemical manufacturing plants, the vestigial remains of the once "Great" Swamp and associated streams collected and accumulated this cocktail of industrial contaminants. One of the abandoned clay pits, itself dug out from the former wetland, became a convenient place to dump the city's trash. With time this landfill became one of the highest elevations in the city, eventually to be reclaimed from its industrial past as a public park representing about a quarter of all the greenspace in Cambridge. And during the later half of the twentieth century, the area accumulated that bric-a-brac of "culture"—strip malls, drive-in theaters, night clubs, motels, used car sales lots, low income housing, etc.—that are often thought best to marginalize to the outskirts of our cities. Finally, despite having some of the most progressive wetland protection laws in the world, the State decided it was in the best interests of its collected citizenry to torture the landscape even further by gouging out additional pieces of the wetland for construction of an end-of-the-line subway station and associated garage as well as to cut another gash in the wetland to expand a regional highway system.

Gazing out the window at what was once the ecologically vibrant Great Swamp, I watch the runoff coursing off the impervious surfaces from the storm of a few hours before, and smell the toilet waste discharged through the antiquated combined sewer overflow (CSO) system that characterizes many old cities such as Cambridge. And I think about how the nature of the all-but-dead wetland continues to force us to reflect upon its former namesake greatness. It was only a few years ago, for example, that the "storm of the century" flooded hundreds of homes when the wetland suddenly remembered its old boundary and the river its old course. Fish, some natives but unfortunately mostly exotics, were seen swimming on golf course fairways, birds paddled around immersed automobiles left stranded in parking lots, and plants continued to sprout from amongst the broken pavement in abandoned lots for months afterward. Nature obviously wanted to come back; but if its return was to be more than ephemeral, it needed guiding and stewardship by concerned humans.

Just as the Great Swamp is a striking lesson in ungratefulness, remorse and shame about how we have treated the wetlands that helped shape this country, the area is now on the cusp of becoming a signature example of dedication, compensation and hope in how to "restore" natural structures and functions in dense urban cores. "Restore" here is in quotation marks because any ameliorative and reparative actions taken towards returning a sense of naturalness to the area and a sense of wildness to its human inhabitants will never be an exact replica of both the ecological and ecocultural relationships that once existed in the area in pre-(European)contact times. None of the enthusiastic participants involved in the reparation—pointedly few of whom are either environmental philosophers or conservation biologists—are not, however, in the least bit bothered by this absence of fidelity to the past. For them—the citizen residents (encompassing local wisdom) and the government officials (embodying contextural knowledge)—it is the *process* of restoration, not only holding the line against a further loss of nature but actually reconstructing snippets of naturalness here and there, that is most important. Whether the final product of their collective efforts mimics that which existed before any of them were born is immaterial to their aspirations.

A team of ecologists and landscape architects have recently completed a comprehensive master plan for the largest remnant of the Great Swamp. Although only 50-acres in size, this reservation represents the largest tract of "urban wild" within the greater Boston area, and has been the central focus for a concerned and vocal group of citizen naturalists over the years. Their dedicated stewardship and publicity efforts have proven successful for the continued protection of this last sliver of the wetland's former glory. Preservation alone, however, was considered insufficient to ensure the maintenance of a natural system of suitable ecological integrity. General plans and design instructions were therefore developed for the restoration processes deemed necessary for the reservation, such as, for example, invasive plant removal, natural channel design, wetland construction, stream "daylighting," habitat creation, aquatic dredging, and native replanting. Because of the importance of urban wilds in blurring distinctions between nature and culture, city and wilderness, and in fostering ecological literacy to inspire urbanites to preserve more natural landscapes located elsewhere, the master plan also devoted much attention to restoring ways in which humans could intimately and deeply experience the landscape rather than simply ignoring it as they had been doing for decades. To this end, the new master plan for the preserved and restored urban wild, championing its overall message of "connectivity," looked upon ways to artistically connect humans to the urban ecology beside them. By revealing this nascent ecology and the relationship of humans to it through the purposeful design of the reservation, balancing concerns about wildlife refugia and public access areas, a mutual healing process was begun.

At the same time, a team of hydrologists and community planners initiated a cultural restorative redevelopment plan to create more favorable living conditions for humans residing in this "edge city" region, and to repair the defective system of landscape plumbing existing there. Accomplishing both of these objectives will support the ongoing restoration efforts of the adjacent nature reservation. The union of the natural and cultural restoration projects will develop through implementation of the watershed strategy of LID or "low impact development." Here, by retrofitting human structures with water sensitive and sustaining plans and designs, all attempts will be made, as much as is feasibly possible, to restore the pre-development hydrologic conditions of the area. The present sea of impervious asphalt and roofs from whence rain runoff is conveyed away into the receiving wetlands of the reservation as quickly as possible, will be transformed through application of LID. Rainfall will now be absorbed by such infiltration devices as parking lot bioretention swales, rain gardens, downspout planters and barrels, and green ecoroofs, with a corresponding overall reduction in the erosion-producing floodwater surges that enter the nature reservation. Further, the direction of runoff into these infiltration areas will improve the quality of the water by lessening of the frequency of CSO events as well as enabling phytoremediation to remove contaminants.

The potential daylighting of a portion of the flow from a buried stream now carried within a culvert is another restoration opportunity. Here, however, the goal is more psychological than ecological. By using the resurrected stream as the spine for a new pedestrian parkway, residents and visitors, most of whom presently have no idea that an historically significant river flows under their feet, will be made aware of its hidden nature. There is no underestimating the importance of such projects for restoring human relationships with, and the subsequent protection of, nature. For one will not protect what one does not love, and in turn, one will not love what one does not know exists. With knowledge gained from restoring the stream through daylighting comes an experience with nature, and through that, an impetus for protection. Enlivening streams in this way is nothing short of enlivening life itself, both human and non-human.

Although I had assembled the team and participated in the master plan for the nature reservation, had introduced the LID concepts to, and been placed on a advisory retainer for, the city public works and community development departments, and had supervised numerous student projects over the years which generated conceptual restorative designs for the area, I couldn't help but think about what was missing. Restoration after all, especially that undertaken within one's own neighborhood, has to be more than an intellectual exercise. There is need to combine the restoration of *both* ecological consciousness and of ecological space. An affair of mind must become an affair of the heart. And the way to effect this transformation is through direct experiential participation in the process of restoration.

Consequently, reflecting on how restoration sometimes involves the destruction of unwelcome elements, I swung my sledgehammer at the garage wall, the bricks of which, given the age of the building, could very well have originated from the very wetland that once existed under my property. Then, disconnecting the rooftop downspout from plunging through the floor and directly into the underlying CSO system, I redirected the pipe out the new hole in the wall and into a rain barrel feeding a small rain garden, thereby restoring a semblance of the natural site hydrology. Also, surreptitiously venturing out into the neighborhood late at night (due to concerns about the legality of the act), I hurled my used Christmas tree off a small bridge into the little stream below. The large woody debris of the tree as it broke down would provide both structural habitat and a feeding platform for instream inhabitants. The location selected for this act—a seemingly dead reach of stream constrained within a concrete straightjacket, devoid of any wood, and running beside a cemetery—couldn't have been more in need of such restoration. Even more important though, was the personal psychological effect of this simple restorative act. For it was undertaken as a form of ritualized performance with the tree given as a gift to the water at preciously midnight on New Year's Eve.

In the end, environmental restoration is about restoring a personal relationship with the natural world through both conscious thought and physical action. Indeed, reparation of the relationship-process of physical action is essential in order for the healing of the nature-product of conscious thought to be sustained. In this respect, it is only through such a bipartite effort that restoration becomes a proactive vehicle for future hope rather than a nostalgic mirror for things past.

Introduction

Engaging Nature and Establishing Relationships Through Restoration Design

Robert France

Restoring Environmental Integrity and Healing Our "World of Wounds"

"Community is perhaps our greatest hope. From community can come the capacity for deep ecumenism, which can release the collective wisdom and passion we need to heal the Earth and ourselves."

—Deroches, L., 1988, Love of Earth and enemy, *In* France, R.L., *Gaian Integrity: A Clarion Precept for Global Preservation,* 1992

Contrary to the doomsayers, nature is not so much at an "end" (McKibben 1990) or "dead" (Merchant 1989), as it is "ill" (Williams 2001), the "damaged world" certainly salvageable, although in dire need of "repair" (Hobbs et al. 1993). Surprisingly, despite a half dozen or so books on various theoretical aspects of environmental restoration, none have really elaborated on the important fact that restoration is actually part of both United States and Canadian law, having been so in some cases for over three decades. Section 304 of the 1972 U.S. Federal Water Pollution Control Act Amendments, for example, specifically states that all citizens must "restore, maintain and protect the integrity of the nation's water and water resources" (U.S. EPA 1975), though just exactly what was meant by the opaque word "integrity" has been an issue of much discussion both at the time (Regier and France 1988) and ever since (e.g. Westra and Lemons 1995).

Often credited to the pioneer American conservationist Aldo Leopold's (1949) brilliant and succinct distillation of his land ethic: "A thing is right when it tends to preserve the integrity, stability, and beauty of the biotic community. It is wrong when it tends otherwise," the word "integrity" occurs earlier in the writings of the poet Robinson Jeffers with reference to the quality of human – nature relationships (France 1992): "Then what is the answer? – Not to be deluded by dreams./ To know that great civilizations have broken down…many times before./… A severed hand/ Is an ugly thing, and man deserved from the earth and stars…/ Often appears atrociously ugly. Integrity is wholeness, the greatest beauty is/ Organic wholeness, the wholeness of life and things, the divine beauty of the universe. Love that, not man/ Apart from that, or else you will share man's pitiful confusions, or drown in despair when his days darken."

Today, theoretical ecologists concerned with the patterns and processes of ecosystems consider biological integrity to be a shorthand expression for the wholeness, diversity, and some would say, corresponding "health," of natural communities (France 2006a). In the latter regard, some environmental managers have gone as far as to construct and use various indices of biotic integrity as biomonitors of environmental degradation and recovery (e.g. Karr and Chu 1999). The implicit precept behind much of this work is a desire to restore, as much as is possible, all species still extant that evolved together in a particular biota.

Ecological literacy—education about the environmental state of the world and the actions needed to fix its many problems (Orr 1992)—unfortunately does not always lead to ecological stewardship (France 2005). Leopold (1949), too, knew that there was sometimes a price to pay for that process of education: "One of the penalties of an ecological education is that one lives alone in a world of wounds." Community-based environmental restoration—the reparation of natural systems and the relationships of humans to them and to each other (Jordan 2003)—is offered as a positive means in which to escape from the lonely apathy resulting from succumbing to Leopold's pessimism.

Although certainly not without discussion and debate in the restoration literature, there is a general belief that integral ecological systems are healthy ecological systems (Steedman 2005). The concept of the healing *by* nature has a rich literature. The solace and therapeutic benefits offered by immersion into nature has its roots in Romantic tourism, Asian garden design, healing spas, and wilderness escape (e.g. Wordsworth et al. 1987; Takei and Keane 2001; France 2003; Colegate 2002). Today, contemporary nature writers, landscape architects, biogeochemists and New Age adherents all follow in this tradition (e.g. France 2003; Marcus 1999; Tyson 1998; Cohen 1997; Grossman 1998; Chard 1999; Lovelock 1991). In turn, the reciprocal healing *of* nature through ecological restoration brought about by the planning and design of landscapes (McHarg 1998; Higgs 2003) offers both community-wide psychological as well as economic benefits (Rozak et al. 1995; Clinebell 1996; Cunningham 2002).

The Nature and Culture of Ecological Restoration

Three important compendia of papers about the theoretical basis of ecological restoration have been published: *Beyond Preservation: Restoring and Inventing Landscapes,* edited by A. D. Baldwin, J. De Luce and C. Fletsch (1994); *Restoring Nature: Perspectives from the Social Sciences and Humanities,* edited by P. H. Gobster and R. B. Hull (2000); and *Environmental Restoration: Ethics, Theory, and Practice,* edited by W. Throop (2000). The first book had its origin in a series of responses by professionals to keynote papers by restorationists W. Jordan and F. Turner. Much of this volume involves a debate as to whether the restoration of damaged nature can either indirectly foster or directly provide as valid a form of environment protection compared to preserving tracts of nature free from human intervention. The second book is based on a conference motivated by the "Chicago restoration controversy," an intriguing paradox wherein mature trees in a suburb would have to be removed in order to "restore" a prairie that once existed at that location in the past. Much of this volume deals with a debate as to whether ecological restoration projects provide social and environmental benefits or whether they are mere fakes that do little more than yet further alienate humans from nature. The third book contains a selection of papers (several from the previous restoration volumes and others from journals) that continues the debate as to the naturalness or artificiality of restoration and as to whether restoration represents either a new paradigm for human – nature relationships based on reciprocity or rather is just another excuse and avenue for the unilateral dominance of nature by well-meaning but naive humans.

The commonality shared by all three books is the overall impression they leave of restoration as being some sort of contact sport involving a good deal of intellectual head-butting. Because it is not the intent of the present book to reiterate these various (and certainly important) debates, I provided a distillation of the various arguments raised in the previous volumes (see Appendix for the essay "The Muddy, Messy Means and Mores of 'Restoring' a Broken World: A Literature Review") to the current group of contributing authors, some of whom were specifically selected to provide new opinions on restoration theory, and thus may not have been completely familiar with the previous literature on the subject. I thought that this strategy would be a way to allow the current group of authors to, in a sense, hit the conceptual ground with their minds running, thus allowing them to explore new directions, or if they did return to the old debates, to do so from fresh perspectives. In short, I wanted to establish a common foundation from which the authors of the present volume could spring off into new directions. Likewise, for readers of this present volume, some of whom may also be exploring these issues for the first time, establishing such a foundation will give a better perspective in which to place and judge the new contributions which follow. This review of material compiled in the Appendix from Baldwin et al. (1994), Gobster and Hull (2000) and Throop (2000) is simply categorized by whether the previous contributors thought restoration to be in the end either a "good" or a "bad" thing.

Two recent books by the leading theoreticians further expand the discussion of the nature and culture of environmental restoration. These other works share much in common with many of the theses explored by the authors within the present pages. In *The Sunflower Forest: Ecological Restoration and the New Communion with Nature,* W. R. Jordan (2003) advances the paradigm of ecological restoration as a reciprocal relationship binding humans to nature through ritual in a shared destiny of hope. E. Higgs, in *Nature by Design: People, Natural Processes, and Ecological Restoration* (2003), reviews the history and development of ecological restoration as a healing relationship that links the past, present, and future in the form of community design. A few of the elements from these important books pertinent to these themes of ritual and design are also summarized in the Appendix.

Restoration Engagement, Relationship, and Design: The Paradigm Put Forward

The chapters in the present volume expand into new directions from the previously documented theoretical musings about restoration. There are two cardinal precepts that underlie and unite all the new perspectives presented herein.

First, all share an overt honesty that not only acknowledges but also champions the role of humans in environmental restoration. And, second, all firmly believe that the *process* of restoration may be just as or even more important than the products ensuing from those efforts; i.e. the concept of restoration as a verb superceding in significance that of restoration as a noun.

These new restoration perspectives can be grouped into three major subject categories that encompass and structure the message of the present book: (1) Engagement, (2) Relationships, and (3) Design.

(1) ***Restoration as a means to engage nature,*** either physically as an act of ceremony and ritual (Chapter 1 – Jordan and Turner) or as pageantry and performance (Chapter 2 – Dannenhauser), or conceptually as a positive act of mutualism and reciprocity with nature (Chapter 3 – Conn and Conn) or potentially as a negative act reinforcing our dominance over nature (Chapter 4 – Kidner).

(2) ***Restoration as a way of establishing relationships,*** such as that of a human healer to the healed landscape (Chapter 5 – Light), that of a resident's life to the inhabited re-naturalized home (Chapter 6 – Mills), and/or that of restorative individuals to the temporal lineage of people and place (Chapter 7 – Spelman).

(3) ***Restoration as a design process,*** with humans expressing their artistic creativity in nature (Chapter 8 – Brown) as a vehicle for cultural awareness and education (Chapter 9 – Collins), or exercising their impulse to reconfigure the landscape (Chapter 10 – Mozingo) as part of defining their relationship with it (Chapter 11 – Ryan).

Together, these contributions advance what might be called a paradigm of "restoration design" (as distinct, for example, from "restoration ecology"). In restoration design then, reparation-minded individuals directly and creatively find ways in which to engage nature by establishing deeper physical and intellectual relationships with their world during the process of re-imaging, reconfiguring, and "restoring" the ecological spaces where they live and their consciousness in terms of how they live.

Before the particulars of each chapter can be summarized and situated in the larger field of environmental thought, it is necessary to explain how the present book came about, how this collection of perspectives is different from those of previous restoration volumes, and what the authors were particularly asked to contribute.

Restoring and Remediating Degraded Landscapes: The Gathering

A conference, an exhibition, a series of workshops, and a symposium were convened at the Graduate School of Design at Harvard University. Titled "Brown Fields and Gray Waters," this conference and associated meetings were sponsored by the Center for Technology and Environment at the Department of Landscape Architecture and brought together more than 50 presenters from North America, Europe and the Middle East. The purpose of the gathering was to focus on procedures for achieving environmental integrity through remediating and restoring degraded terrestrial and aquatic landscapes. There were three specific goals: first, to present promising options and limitations in the cleansing of polluted brownfield and industrial sites and the repair of wetlands and buried streams; second, to introduce new restoration and remediation approaches of particular interest to professionals, academics and researchers working productively in the fields of design, ecology, engineering and technology; and third, to examine technical studies and key information and the interdependence between innovative site technologies and novel planning and design strategies and processes.

The gathering was perhaps most successful and instrumental in demonstrating the extremely broad range of interests and professional approaches encompassed within environmental restoration and remediation. In addition to landscape architecture, the following disciplines were represented by the presenters: land-use planning, environmental engineering, conservation biology, real estate development, environmental law, public health, politics, hydrology, urban design, government regulation, watershed management, environmental education, soil science, and community management. The gathering was attended by over 200 practitioners, academics, students, and interested private citizens. Topics covered in the conference and the professional development workshops (the latter titled "Building a Restoration Toolbox") included the following techniques: waste disposal, end-use park design, urban renewal and planning, wetland restorative engineering, stream daylighting, mining reclamation, natural river channel design, industrial landscape restoration, phytoremediation, bioengineering, soil reconstruction, thermal remediation, groundwater monitoring, and restoration ecology; and the following non-structural elements: community development, health and legal aspects, environmental toxicity, international exchanges, government regulation, financing and real estate, education and ecotourism, environmental justice, and technology transfer; all undertaken in the following locations: landfills, brownfields, post-industrial factories, abandoned mines, wetlands, former military bases, former gas manufacturing plants, rivers, buried steams, and degraded urban cores. These case studies are being collated and will subsequently be published in a volume titled *Handbook of Regenerative*

Landscape Design (France, R. (Ed.), in prep, CRC Press).

In addition, an exhibition, titled "Reclaimed! Case Studies of Reclamation Processes and Design Practices," was opened at the conference and ran for two months afterwards. Five of the most complex, challenging, educational, important and inspirational projects from around the world were presented: Xochimilco in Mexico City, Clark County Wetlands Park in Las Vegas (France 2009), Westergasfabrick in Amsterdam, Fresh Kills landfill in New York City, and the Wetland Centre in London. These studies were specifically selected in order to promote the message that the most important restoration and remediation projects cannot be undertaken without the cooperation and integration of many different disciplines. In other words, restoration is both far too important and far too complex to remain the purview of any single group of practitioners, such as for example, ecologists.

Prior to the conference, workshops, and exhibition—most of which dealt with the more technical aspects of restoration and remediation—a half-day symposium, open to all conference attendees was convened. Titled "Healing Natures, Repairing Relationships: Landscape Architecture and the Restoration of Ecological Spaces and Consciousness," the symposium addressed questions of "why" rather than "how," "what," "where," or "when." In addition to a handful of plenary lectures at this pre-conference symposium, a group of specially invited individuals, of whom most did not have opportunity to speak due to time constraints, were asked to sit through the subsequent conference and to talk to as many of the participants and attendees as possible in order to gauge the philosophical zeitgeist of the theory of environmental restoration and remediation. Following the Brown Fields and Gray Waters conference, this group of "conceptual spies" convened in a private meeting to share their impressions and knowledge gleamed from conversations and listening to conference presentations, and to begin planning their own contributions to the present volume.

Healing Natures, Repairing Relationships: The Book

The eleven chapters in the present compilation are from 13 authors, 8 of whom have not previously contributed to the literature on ecological restoration theory. All these new individuals have, however, explored various aspects of relationships between culture and nature in their own writings or activities, and were purposely selected because of this. In addition to environmental philosophers and natural scientists (which have generally been the writing staple of the previous restoration compilations), the present book also contains papers by psychologists, landscape architects, ecological artists, and a nature writer. Each author was asked to draw upon her/his own disciplinary background and to adapt this knowledge to the topic of "healing natures, repairing relationships: restoring ecological spaces and consciousness."

A few words of explanation about the title of the book in terms of how this might have directed the authors in their tasks. "Healing" was selected to tie into both the psychological and legalistic aspects of restoration (as described above). "Natures" was purposely selected in its plural state to circumvent much of the endless dithering that has previously both structured and limited the field of restoration theory in terms of whose idea of which precise and "good" nature (in singular) is what "bad" nature being restored to when, and by whom, etc. Also, "nature" of course has just as much to do about humans as it does about non-humans. Indeed, some have gone as far as to suggest that the very concept of nature is itself a human construct shaped or at least influenced by our consciousness (e.g. Evernden 1992; Cronon 1996). "Repairing," *sensu* Spelman (2002), was selected to emphasize the fact that what restorationists are really doing, despite occasional self-aggrandizing prose suggesting otherwise, is simply fixing broken parts (of both lives and of the ecosystems in which those lives are embedded). Although it is seductive to liken the act of ecological restoration to physicians repairing damaged bodies, it might be more appropriate to perhaps envision it akin to the humble reparation work done by plumbers. In short, though restorationsists may be proud of their efforts in repairing damaged ecosystems, caution is needed to prevent hubris replacing humility, for it should never be forgotten that restoration repair, like recycling, is something done *after* the sins have been committed. "Relationships" was selected (again in plural) to stress the important tenet made previously in the literature and reiterated repeatedly in these pages, that "all truly sustainable, and therefore successful, environmental restoration projects are as much about restoring degraded human – nature relationships as they are about simply repairing degraded physical landscapes" (France 2001). "Restoring" was chosen as a convenient umbrella term for the whole suite of possibly more accurate expressions of intent, such as, for example: remediating, reclaiming, rehabilitating, reinhabitating, renovating, renewing, rejuvenating, regenerating, reconstructing, recreating, repairing, recovering, redesigning, etc. What all these words have in common is the prefix "re," indicating a return to some preferred state. Readers should therefore interpret the word "restoration" in what follows as nothing more than a broad concept and not to infer that the authors are necessarily implying a return to some precise state of imagined ecological naturalness that existed presently. "Spaces" was purposely selected over "places," even though the former is a less accurate term for describing an inhabited and loved physical landscape (Tuan 1977); this is simply because the word "spaces" can also refer to conceptual landscapes just as much as to physical ones. And finally, "consciousness" was chosen for just the same reason; i.e. what we are most interested in this book is in examining how restored ecosystems have been either shaped by or may in turn influence human thought processes, rather than in the technical reassembly of the broken bits and pieces of non-human nature itself.

Restoration Engagement

In their chapter, William Jordan and Alex Turner herald the benefits of restoration as a direct means of participation in ecological processes in a way that forces individuals to confront their past, present and future relationships with nature. Through the union of environmental thinking and tinkering, restored landscapes can be looked upon as hybrid creations of culture and nature. Jordan and Turner rightly acknowledge that although such a union may be deeply troubling to many, the creative act of restoration, which may involve some destruction of existing natural systems, is nevertheless of overall benefit to how we structure our relationship to the natural world. In this regard, the greatest benefit of restoration may very well be in its influence on human consciousness rather than on natural system repair. Ecological restoration, Jordan and Turner believe, is therefore not just a technical exercise, but instead should be looked upon as a ceremonial performing act of ritual—a sort of anthropological passage towards maturity.

As Mark Dannenhauer states in his chapter, restoring brownfields involves a co-evolution of nature and culture rather than simply a process of ecological succession characterizing the hands-off approach often used in greenfield restoration. Ritual emerges from ceremony, and performance creates a living history which can be used to successfully bridge between social and ecological systems. Drawing on the rich heritage of traditional sources from around the world, Dannenhauer promotes a technique of mapping participation on restored landscapes. His strategy is to derive a method whereby performance is used as a biomonitor to judge the health of the environment and to gauge the effectiveness of restoration efforts. In this respect, as much attention needs to be focused on the success of restoration in terms of people's relationships to the landscape as on the physicality of the reparation work itself. Dannenhauer's belief is that in true restoration and healing, cultural diversity will parallel biological diversity.

In their chapter, Lane and Sarah Conn address the forced dichotomy between physical and human nature, and advance the idea that the environmental crisis is due to a crippled consciousness. Ecological and human health are intertwined, the field of ecopsychology being the examination of the expanded identity of our larger selves. We need to experience a mutual relationship with nature, the Conns state, if we are to hope to restore our ecological consciousness. Further, the inner nature of that relationship with outer nature will influence the form and success of our restoration efforts. If restoration is approached as just another form of dominance tinkering it will produce landscapes that are projections of our own inner hubris rather than arising from our perceptions of outer nature. It is important to engage restoration just like any other relationship as a dance of reciprocity. By experiencing our embeddedness in nature, by acting with our hearts as much as with our hands and heads, the Conns believe that we can make ecological restoration an act of healing for both the outside and inside worlds, each in dire need of such repair.

Because we shape the world to fit our preconceptions of how we imagine it should be, David Kidner cautions in his chapter, of the danger in letting restoration become a cognitive extension of the very processes that caused the damage in the first place. It is possible for restoration to be used unscrupulously as part of the destructive assimilation of nature, as an excuse to continue business-as-normal actions infused with the belief that no matter how much harm we can inflict upon the land there will always be ways to patch things up and make everything as good as new. To have restoration be supportive rather than destructive of wild nature, Kidner believes it critical that we need to *feel* not just think restoration. Restoration therefore needs to be based on the world itself, not merely our idea of it, and it should be approached with humility not hubris. Restoration is about remembering, Kidner also tells us. We must, therefore, be aware that history, like nature, is a dynamic process, impossible to characterize as a static entity frozen to any single moment in time.

Restoration Relationships

Because of our estrangement from nature, it is more important, Andrew Light argues in his chapter, to find ways to reconnect people to the nature in their own backyards than to that within distant wilderness areas. In short, nature is not something excluded from culture, but rather is something that can be experienced right at home. Light reviews the ongoing debate about the worthiness of restoration as a means for fostering environmental consciousness, believing that the best way to defend the land from rampant development is to directly engage in a restorative relationship with it. And it is the nature of the relationship that is just as important as is the relationship to nature. Restored landscapes possess embodied history, their true value being a function of our relationship to them that transcends the mere physicality of their component bits and pieces. Whether a restored nature is a realistic or an artifactual structural replacement for that which has been lost—in short, whether the restored landscape is a part of, or apart from, physical nature—is, in the end, of secondary importance compared to the participation of those involved in its creation. For it is in the mirrored relationship of restored natures to the participants, the extension of selves in the work done, that Light believes to be the most significant result arising from any restoration project. For by maximizing the public participation in restoration, the likelihood of sustaining success will increase due to creation of social capital.

The chapter by Stephanie Mills is significant in its implicit message that just as it may be deemed important to restore an exploited landscape, it is equally as important to "re-story" that place through the telling of a personal recounting. And the way to do this is through reinhabitation, which may be regarded as the true end result of restoration/remediation and essential for its long-term success. Humans, Mills argues, need to find ways to live harmoniously within a restored nature, witnessing and learning from the industrial history of the site at the same time as recognizing that health is an inherent right of all landscapes, past, present, and future. Whereas restoration provides the balance that civilization needs in order to affect this, it is reinhabitation that in the end provides the cure for our estranged relationship with the natural world. The act of living with humility and sensitivity in a renaturalized landscape, in a sense making the restoration work sustainable, is, Mills acknowledges, a long process, requiring both patience and dedication.

In her chapter, Elizabeth Spelman examines restoration as part of the larger enterprise of reparation. From broken objects through broken relationships to broken concepts, there is much in our world in dire need of fixing. By tending and mending our wounded world, it is human nature, Spelman informs us, to attempt to react against and possibly triumph over time's inevitability. While the act of reparation is to acknowledge the inherent fragility of our world, there is no escaping the fact that by allowing the past to exist into the future, reparation is also an exercise in presumptuousness. For as Spelman cautions, repair also carries with it an inherent hubris in which a decision is made to give greater value to one instance of time over that of another. Herein lies the rub implicit in all acts of restoration: by repairing something through returning it to an earlier state of its past, we are erasing the past process of deterioration, which, it could be argued, is just as much an element of an object's history as is anything else. There is a danger too, Spelman posits, in losing track of the reality of the thing being repaired in terms of just how much is original and restored and how much is new and has been recreated. We may need to adopt a different form of aesthetic wherein we develop an appreciation for the greater beauty in imperfect partial objects than in embellished wholes. How would such a concept of embracing brokenness, establishing a sort of "romance of the ruins," affect the theory of restoration?

Restoration Design

By giving attention to history and to creating a sense of place, ecological art, Jill Brown describes in her chapter, seeks to revitalize landscapes rather than merely beautifying them. Ecological art incorporates metaphor, instills a concept of perseverance and hope, and develops an aesthetics of survival, all wrapped up together in its major role of societal education. Brown shows that through ecological art projects that blend aesthetics and science, some of which being purposely subversive in their execution, nature can be both ecologically healed and sociologically reclaimed. The concept of functional art, wherein human-made installations foster biodiversity or stabilize eroded shorelines, for example, or where human arrangements of plants bioaccumulate and thus "mine" sites of their toxic burden of chemicals, provide an interdisciplinary lesson in how culture and nature truly become one in the act of restoration writ large.

As Tim Collins outlines in his chapter, ecological art, like restoration ecology, fuses nature with culture as a participatory and healing enterprise. By creating an environmental aesthetic, eco-art can act as an agent of change based on direct subjective experience rather than treating nature as merely an object for distant artistic contemplation. Environmental aesthetics, Collins instructs, unifies nature and culture until there is no difference, towards which eco-artists work as witnesses, advocates, or activists. It is not so much that preservation is wrong, Collins believes, as it is simply outdated for the twenty-first century. Eco-art can be one of the most valuable tools we have for bringing acts of restoring nature from the conceptual dark ages into a contemporary zeitgeist, in a sense, restoring the purpose of restoration that of connecting and healing culture as well as connecting and healing nature. And by revaluing brownfields, supporting biodiversity, dealing with urban infrastructure, reclaiming polluted lands, and confronting issues of environmental justice, Collins demonstrates that eco-art may be the most truly interdisciplinary form of "ecohumanist" restoration that we have.

In her chapter, Louise Mozingo reminds us that for more than a century landscape architects have been creating places by fusing and contrasting cultural and ecological systems in the restoration of both natural systems and human - nature relationships. Whereas other disciplines continue to debate and redebate nature vs. culture, natural vs. artificial, preservation vs. restoration, etc., landscape architects build iconic and contrived "second natures" such as Boston's Back Bay Fens which produce far-reaching and lasting social and environmental benefits. It is important, Mozingo believes, for restoration to pay more attention to urban environments because of their ability to foster ecological literacy amongst the greatest number of people. Inspiring people to protect nature in remote, more pristine locations will develop from their relationships toward restored nature within cities. And it is here where landscape architects can play their important role in making the ecological processes of reparation overt through their revelatory designs. There is an important need, Mozingo instructs, however, for restorative landscape architecture to move beyond simple education toward true reparative actions upon the land. And there is much we can continue to learn about the practice of healing nature and repairing our relationship to it from the important work of such pioneers as Frederic Law Olmsted.

Despite the important historic and present roles of landscape architects in environmental restoration and remediation, the bulk of publications have been by scientists and engineers. Robert Ryan's chapter examines results from a survey concerning attitudes of various disciplines toward restoration that he conducted at the Harvard conference. In contrast to ecologists, landscape architects focus on more degraded and often urban environments where production of the intentionally designed landscapes is considered to not only be acceptable but sometimes actually desirable. In order to heal the land, Ryan's survey showed, it was not thought important to erase all evidence of human presence, both beneficial or detrimental, to the environment; i.e. honestly acknowledging the history of a site may be more significant for the lessons that can be learned about human relationships with nature than from the purely ecological benefits offered by sites restored to a state that attempts to disguise their history. By creating opportunities for experiencing restored nature, landscape architects, Ryan found, focus on human communities as much as they do on non-human environments. In this respect, restorative landscape architects pursue a much more holistic approach to restoration than do restoration ecologists. Rather than working toward and fretting about imitating some idealized past state of pre-contact nature, landscape architects believe their primary purpose is to first, heal degraded landscapes, second, to stop further environmental abuses that caused the situation, and third, to reconnect people to the improved land. Ryan believes that ecological restoration for landscape architects is just one piece in how they can help to develop a larger world view about human relationships with nature.

Conclusion

Although the perspectives pursued in the present book arise from different disciplines and occasionally from divergent attitudes, taken together they contribute to the coherent conclusion that the emerging field of restoration design is that process of environmental reparation and rehabilitation conducted squarely within the blurred interface between nature and culture. These contributions thus serve to circumvent previous theoretical debates that have plagued much of the ecological restoration literature and rather advance the notion that the design of that which takes place upon the landscape is fully part of the same process as that which takes place between the ears. And it is the process of engaging such a bipartite design through relationship building that is the key to developing this deep, ecumenical form of restoration.

Acknowledgements

I would like to thank the authors whose thought-provoking papers are included in this collection, Clay Morgan and others for their critical reviews of early manuscript drafts, as well as all those individuals involved with helping to organize and run the Harvard restoration conference from which the present effort was derived. The conference was sponsored by the Design School through Dean Peter Rowe and Landscape Architecture Chairman George Hargreaves, and was co-facilitated by Niall Kirkwood. Production of this book was made possible by a William F. Milton Fund and a Tozier Fund from Harvard University.

Part 1

Engagement

Chapter 1

Ecological Restoration and the Uncomfortable, Beautiful Middle Ground

William R. Jordan III and Alex Turner

In this chapter we explore the idea that the practice of ecological restoration provides the basis for a new kind of environmentalism, one that offers the possibility of a truly healthy relationship between our own species and the rest of nature, exemplified by the "natural"—or classic—landscapes that are commonly the models for restoration projects.

We believe that this is true partly because restoration provides, as our other major paradigms of relationship with the "natural" landscape have not, the basis for a relationship with these landscapes that is both active, allowing us to participate in the ecology of the system, and at the same time constructive. It arguably provides, in a way that no other activity we can think of does, an opportunity to engage the full range of our aptitudes—aptitudes for manipulation and invention as well as those for discovery, observation and appreciation—in actual work in such a landscape without compromising its value *as* a "natural" system. In doing so it defines, we might say, an ecological niche for the human in the classic ecosystem. And when successful it leaves the ecosystem, on its own terms, better than it was to begin with.

At the same time we believe that restoration is also valuable, perhaps in an even more important way, for precisely the reasons that several generations of environmentalists have found it problematic—because, despite its promise as a way of reversing environmental damage and establishing a positive, constructive relationship with the natural landscape, it also forces us to confront the most troubling aspects of that relationship.

Restoration, as A. Spirn reminds us, is not a new idea.[1] Even in its most ambitious form—what historian M. Hall and the senior author have recently termed "holistic restoration"—that is, restoration of the whole, with a studied disregard for human interests[2]—restoration dates back at least to the early decades of the last century. Yet for most of the century environmentalism—indeed, a whole succession of environmentalisms—first ignored and then later resisted it.

This, we believe, was a mistake—in fact, one of the defining mistakes of twentieth century environmentalism—and so it is worthwhile to consider what might have been the reasons for it. Some were obviously technical and economic. Environmentalists have been skeptical about restoration because it is often complicated work that practitioners and ecologists don't necessarily know how to do very well, and also because it is often expensive, raising the question of whether it is feasible on a scale commensurate with the environmental damage that humans have caused in recent centuries. Others are political: environmentalists have always been concerned that the promise of restoration might be used—as in fact it has been used—to undermine arguments for the preservation of existing natural areas.

These are serious problems, not to be dismissed lightly. But underneath all of them, we believe, are even more fundamental concerns of a psychological and even spiritual nature. Restoration is, at this level, a deeply problematic business. Like any form of agriculture, it involves killing and discrimination. It involves a measure of hegemony over other species. Whatever our intentions, however sincere our studied disregard for our own interests, restoration does contaminate nature with human intentionality, as critics of restoration like R. Elliott and E. Katz have pointed out.[3] And the results, even when we are reasonably successful, are puzzlingly ambiguous. What is this thing—this prairie or wetland—that we have created? Is it real or, as some have argued, fake? And what kind of thing is it? Nature? Or artifact? Or something troublingly in between?

This is not the way we have expected a healthy relationship with nature to feel. And yet there are good reasons, suggested by classic institutions such as ritual sacrifice, for example, or often harrowing rituals of initiation into community, or the great myths of origin and relationship to believe that this is exactly the way the deepest forms of relationship do feel, or, more accurately, that these feelings are the price we pay for the experience of relationship.

The world, the myths remind us, is a violent place, nature itself being at odds with itself in the very act of creation. And yet, the myths also show, the great, transcendent values such as truth, beauty, meaning, community and the sacred can be our reward for confronting this violence and ambiguity and finding psychologically productive ways of dealing with it. Indeed, the poet and literary critic F. Turner argues that this is exactly what these higher values are—nothing less than the neurological reward evolution has provided for the hard work of confronting the trouble at the heart of creation and inventing ways of dealing with it creatively.[4]

We experience this trouble, Turner argues, most intensely in experiences of relationship and of change (relationship in time), which dramatize the transience, mortality, difference and limit that are inherent in creation and naturally give rise to negative feelings such as fear, regret, anxiety, envy, horror and, most fundamentally, shame. Since these feelings are, arguably, inseparable from our experience of the world, community with the world depends on finding a way past or through these feelings. The trick, then, is not to deny or downplay the trouble and tensions and bad feelings we encounter in the course of activities in which we engage nature—activities such as restoration or agriculture or medicine or the rearing of children—but to confront them and to find ways of passing through them, transmuting, as Turner says, the shame of creation into values such as community, meaning and beauty.

Here, we believe, is the greatest value of restoration, the real frontier in our search for communion with the classic landscapes that are commonly the models for restoration projects. Restoration, like any way of engaging nature is fraught with the uncertainties, incompletions, ambiguities and murders that are, we must recognize, inherent in nature itself, a manifestation of what mythologist Joseph Campbell calls, in an arresting phrase, "the monstrosity of the just so,"[5] and what Turner, in a more playful spirit, calls the amateurishness of creation.

A heightened awareness of this, achieved through the experience of restoration, may be the single most important thing about restoration. And it is very much in evidence in this discussion—a promising step, we believe, toward higher things. Thus J. Beardsley (Harvard conference), while celebrating the creation of a recreational area out of an abandoned industrial site in Germany, reminds us of the danger that successes such as these will encourage us to overlook the destruction that played so large a role in their history. In a similar vein, L. Mozingo (Chapter 10) expresses concern that the success of community-based landscape restoration projects in urban areas may foster a specious optimism that overlooks the destruction that continues elsewhere, in rural or wilderness areas, or in other countries. And D. Kidner (Chapter 4) has described his own well thought out catechism of misgivings regarding restoration, including the concern that it may well be—or may devolve into—just another way of exercising human hegemony over nature, and that the idea of restoration itself may reflect a naive and dangerously innocent idea of the reversibility of ecological and biological time.

These concerns reflect the emotional and existential difficulty that is inseparable from the practice of restoration—and, as we have suggested, from the process of creation itself, in which the restorationist presumes to participate. Especially relevant here are two other observations directing our attention to the psychological difficulties inherent in the practice of restoration.

The first is A. Spirn's observation (Harvard conf.) that landscape designers such as Olmsted and Jensen were among the pioneers in the development of restoration when it began assuming its modern form around the turn of the last century, but that over the years people lost sight of the fact that the landscapes they created—landscapes that, she notes included explicitly wild areas, such as parts of Yosemite National Park, as well as urban icons such as Central Park in New York and the Emerald Necklace in Boston—were in fact designed, planned, deliberately created landscapes—and began to regard them as natural. The same thing happened, as the senior author knows from personal experience, at the University of Wisconsin-Madison Arboretum, itself the direct product of an historic restoration program, which until recently had all but lost psychological contact with its own history as a site of pioneering experiments in restoration. This is interesting because it illustrates the difficulty Americans have had holding onto the idea of restoration, taking it seriously, and putting it to work. And the most fundamental reason for this, we suspect, quite apart from purely practical considerations, is that restoration violates the categories of "nature" and "culture" Westerners have used since ancient times to make sense of the world.

Thus we have an assortment of words such as "garden," "farm," "ranch," "plantation," "park," and so forth to refer to cultivated or otherwise managed landscapes. And we have a perfectly good, if in some quarters hotly contested, word—"wilderness"—to refer to the opposite of this in uncultivated, undisturbed, "pristine" landscapes. But when confronted with a restored landscape, this hybrid or mongrel thing that so decisively violates the categories of nature and culture, we have trouble talking about it and so, of course, hanging onto it conceptually, making sense out of it and keeping it alive in the landscape.[6]

The problem is that, at least in a Western context, restoration falls in a kind of epistemological middle ground where, lacking a secure label, it doesn't really exist as part of an integrated world view. And since such mongrel entities are a source of anxiety, we tend to forget them—and the inevitable result is that they eventually disappear, first from consciousness, and then, as a direct result of that, from the landscape as well.

The second observation we would especially like to draw attention to here is M. Ukeles's observation (Harvard conf.) that American society has lost touch with its garbage—a condition that she regards as a calamity that she likens to a cultural "lobotomy" (it occurs to us that "colostomy" might be a better, or at least complementary metaphor).

Garbage, Ukleles suggests, is vitally important for many reasons, but most deeply because it is inseparable from life, and therefore must somehow be made sacred. To this we would add that it is precisely because it is an occasion for the experience of shame that it offers an opportunity for work like Ukeles, which aims at the creation of transcendent values such as community and the sacred.

Interestingly, she notes that garbage is also a "resource stripped of its identity," an observation that resonates in an interesting way with our inability to "identify" restoration, or to label it in a satisfying way. The restored landscape, perhaps no less than the derelict landscapes in which restoration is carried out, is, at least in this sense, cultural garbage—a kind of thing that violates the categories of "nature" and "culture." Like kitchen garbage, it is also necessary. And so, like kitchen garbage, it offers an opportunity to encounter shame—the shame of the unwanted, the rejected and un-named, the shame of the destruction that accompanies creation and our dependence on that destruction, the shame of what Robert Frost called "the universal cataract of death"[7] on which life itself depends—and so offers an opportunity for communion with nature and the creation of the values that arise from this communion.

In both cases—whether we are talking about garbage or a restored landscape—when we talk about making this metaphysically troubling thing sacred, or making it the occasion for an encounter with the sacred, it is clear that we are talking about an idea of value that is radically different from the one most of us take for granted. In particular, we are very far from Aldo Leopold's complaint, in *A Sand County Almanac,* that an ecological sensibility condemns one to living "alone in a world of wounds."[8] It is a considerable distance from this, ultimately despairing idea (despairing because life is linked not only to death, but to killing), to Ukeles's insistence that the unending torrent of animal and vegetable corpses—manifestations of a nature in which brothers and sisters live by feeding off one another's bodies—that she deals with in her work must somehow be made sacred, and offer, through art, an occasion for the creation of value.

The key issues here are the categories we use to make sense of the world, the conceptual boundaries these entail, the psychological bearings or consequences of these categories and the boundaries that define them, and their relation to the creation of value. Here we are concerned in particular with the categories of "nature" and "culture." Some environmental thinkers, noting that these categories are not shared by all human cultures, have supposed that they underlie the alienation from nature that they take to be a peculiarly Western malady. This, some have suggested, might be corrected, and our environmental difficulties taken out by the roots, if we could replace this set of categories, and the anthropocentric view of the world that arguably accompanies it, with a more egalitarian, eco- or biocentric view of the world based on a different set of categories.

What we argue in this chapter is that, while the categories a culture uses to make sense of the world are obviously an important feature of its worldview, it is not enough to characterize a culture or a society in terms of these categories alone. Of even more fundamental importance is a culture's idea of categories in themselves, and how it handles what we might call the experience of categories—that is, of distinction, difference and discrimination, apartness, classification, identity, exclusion and inclusion—and the attitudes and feelings, especially negative feelings such as shame, that arise from this experience. There are, for example, societies that apparently do not think in terms of "nature" and "culture" as these ideas are understood in the West. Yet these societies have elaborate ways of dealing with the ambiguities and trouble arising from the categories they do use to make sense of the world.[9] There is, it seems, no escape from this existential trouble. And to overlook this, as environmental philosophers characteristically have, is simply to overlook the existential elephant in the room—the facts of experience that define the affective dimensions of any relationship. It is, in other words, to build on and to put into play an idea of categories and boundaries and their relationship to values and value creation that ultimately *precludes* the discovery, creation or experience of value in its highest, or deepest forms. And it is to fail to come to terms with, or even to acknowledge aspects of experience that lie at the heart of environmental problems.

We believe that this failure is endemic to environmental thinking, that it limits the capacity of environmentalism to do the psychological work of self-transformation and community-building central to its mission, and that this fundamental conceptual weakness underlies the limitations of traditional forms of environmentalism recently outlined by M. Shellenberger and T. Nordhaus.0 We also believe that the ideas about value that F. Turner has articulated provide an alternative set of ideas that provides a way—not of solving this "problem," since it is not a problem but a condition of life—but of dealing with it productively. We believe that these ideas have important implications for every aspect of environmental thinking and practice. Our purpose here is to explore their implications for ecological restoration, understood as a context for the creation of values such as meaning, the sacred and—especially—community.

In his 1994 book *Earth's Insights,* environmental philosopher B. Callicott explores a wide array of cultural, religious and philosophical traditions in a search for a set of ideas common to all of them that might serve as the basis for a shared environmental ethic.[11] As a guide or point of reference he uses a cluster of ideas associated with the land ethic of Aldo Leopold, which we may regard in this context as an hypothesis regarding the terms of a healthy relationship between humans and the rest of nature. These include an emphasis on the importance of a non-anthrocentric world view and a kind of egalitarianism among species, including humans, who, Leopold argued, should consider themselves no more than plain members of the land community.

What Callicott finds is that many of the cultural traditions he explored are fundamentally consistent with this set of guiding principles. He notes, however, one major exception in some of the cultures of sub-Saharan Africa. Here, he writes, humans have lived as long as in any place on earth, with less impact on the landscape than in many other places. At the same time he finds that the world views of many African societies are inconsistent with key elements in the cluster of ideas he used as a reference point in his explorations. Specifically, he finds, that the world views of some are explicitly anthrocentric, positing a high god who created the world and the creatures in it for the sake of humans. And, while he notes the sensitivity to the environment reflected in African art and ritual, he thinks it unlikely that these societies will join, much less contribute to, the post-modern land community he envisions.

Callicott calls this the African "paradox." But since his project is, after all, presented as a kind of experiment to test a hypothesis by observation, it would perhaps be more accurate to say that it is simply a very large body of data that violates—and so invalidates—the hypothesis. If a diverse array of African societies have managed to maintain a satisfactory relationship with their surroundings for thousands of years while holding a world view that is in crucial ways inconsistent with a set of criteria, then there must be something wrong with the criteria.

The idea we want to put forward here is that the difficulty Callicott is dealing with lies in the land ethic itself—not in the idea of an ethical imperative based on appeals to key ecological and transcendent values such as community and beauty, but on ideas of those values, what they are and how they are achieved that Leopold, like many other environmental thinkers, apparently took for granted. Callicott has done valuable work updating our ideas about the ecological values of integrity and stability in response to developments in ecological thinking, but those concerned with the land ethic have paid far less attention to the transcendent values of beauty and community. Since, as Callicott has argued, community in particular is the cornerstone of the value-base underlying the land ethic, this is a serious matter, especially for a society such as that of the modern West that is notoriously unfriendly to community, and in which the experience and institutions of community have arguably been losing ground steadily for the past five hundred years.

Aldo Leopold's accomplishment was to take the idea of community that ecology had borrowed from the cultural commons and to reassert in behalf of the biotic community the ethical claims originally associated with the human community. In his discussion of community in "The Land Ethic" and elsewhere, however, Leopold dealt almost exclusively with the ecological aspects of community and with community as an ethical ideal.[12] While he insisted on the importance of love as a precondition of community, he had little to say about the psychology of either love or community, especially in their more troubling aspects. His vision of an ecocentric world in which categories such as "nature" and "culture," or "self" and "other" have either been defined away or have lost all or most of their negative emotional significance is clearly an attractive one. It contrasts sharply, however, with the experience of community that characterizes either our own immediate experience of relationship, or that is reflected in the traditions and practices of societies, from traditional, place-based societies to gangs and secret societies, in which the institutions of community actually predominate or at least are taken seriously.

Consider, for example, the often harrowing rituals by which individuals are initiated into many traditional communities traditions[13]. Or the Sundance of the American Plains Indians, a ritual of communion and world renewal that entails self-inflicted torture. Or the demanding disciplines of self-knowledge, reconciliation and communion characteristic of all the major religious traditions. Or even, extending our consideration to other species, ethologist Konrad Lorenz's idea, based on studies of the mating behavior of birds, that love itself is the product of rituals that serve to dispel the tensions arising from territoriality.[14]

Or consider the institution of ritual sacrifice, found in many cultures worldwide, by which a community acknowledges what F. Turner has called its solidarity in crime with the rest of nature,[15] and thereby attains communion with it—a deliberate killing that calls into question the aspiration toward a world *without* wounds implicit in Leopold' reflection on the cost of an ecological conscience. Such a world, it seems, would also be a world without communion, and so without either community or beauty.

These disciplines and institutions are not *ideas* of community, a value that is, at least in the abstract, unproblematic and easily prescribed, but rather the concrete responses of various societies to the actual experience and practice of community. What they reveal is that community in the traditional or classic sense is not easy but emotionally difficult and costly. It does not just happen, but has to be earned by the hard work it takes to negotiate the boundaries between the self and the other. These boundaries are apparently deeply problematic, but they are *also*, Turner argues, the occasion for the creation of values such as community and beauty that are the basis for ethics. This being the case, where a culture draws these lines—whether between humans and other species or in some other way—is of far less importance than what it *does* with them, and does specifically with the shame they generate. And if this is true, then community is achieved not by downplaying the boundaries between self and other, or by shifting them or defining them away, but by acknowledging them, confronting them and finding ways to deal with them productively.

To deal with this, however, it will be necessary to step outside the bounds of philosophy to explore the experience of relationship from the perspectives provided by disciplines such as anthropology and theology—not in a search for ideas, but in order to investigate those technologies of the imagination that cultures use to transcend the categories inherent in language and ideas. These are, of course, precisely those technologies of art and myth, symbol and ritual that Callicott noted as a possible basis for an environmental ethic in African societies. These, we may note, have not traditionally been the business of philosophy, but they are very much the business of an artist like Ukeles, and it is important to emphasize that these both underlie and transcend not only ethics, but any categories in which an ethic might be framed. This is the domain of what the British American anthropologist V. Turner called "communitas," the experience of community, which is achieved through ritual, and which he saw as the "primal ground of social structure."[16] It is also the domain of what theologian C. Pickstock has called "the liturgical consummation of philosophy"[17]—the process by which humans (and, Lorenz would point out, also geese and many other species as well) come to terms with the tensions that arise, not from the ultimate unity of the world, but from its manifest and troubling diversity and otherness—not, it may be, in the sense of being radically other, but simply of being an-other.

Viewed from this perspective, African anthrocentrism is important, not as a puzzling exception to an environmentally more-friendly ecocentrism, but as a forthright acknowledgment of the most troubling aspects of difference, including the natural self-centeredness that is necessary for the survival of the individual or the species, and the violent hegemony of a heterotrophic species over those species that serve as its prey.

This, together with the example of institutions that deal with the shame arising from these differences in a creative, productive way, we believe may be the great contribution African (and other pre- or non-modern cultures have to make to environmental thinking. But this is not peculiar to African traditions. In fact, it is an element of all the traditional and classic cultures Callicott explores in his book. To test this idea, it will be helpful to return to the "data" on the anthrocentric cultures Callicott describes to determine whether we can find evidence that such cultures apprehend and confront the negative aspects of difference, relationship and change, and that they provide ways of dealing with it effectively, not in terms of an abstract ethic, but through the more concrete forms of cultural expression found in myth, symbol, ritual, art and religion.

Aldo Leopold wrote that we must find the "key log" that has to be removed to release the regard for other species that has developed in the course of evolution[18]—an idea that anticipates the more recently developed idea of biophilia.[19] The urgency we sense here is appealing, but the metaphor may be misleading, suggesting as it does that communion is natural, easy or given, like the flow of a river—that for some reason, our relationship with other species, or with "nature" in general, can or ought to be less problematic, less troubled or emotionally challenging than our relationships with members of our own species.

What we find on close examination of African cultures, however, is unmistakable evidence that community—or communitas—when they exist, are not at all the result of a release of natural energy—or at any rate not that alone—but rather the result of deliberate cultural work, carried out through ritual. Rituals of divination, revelation, initiation, possession, and the healing of affliction are inseparable from the cultural lives of traditional African peoples, and V. Turner, who worked with the Ndembu people of what is now Zambia during the 1950s, wrote in his classic book *The Forest of Symbols*, that he came to see these rituals as a process "whereby groups became adjusted to internal changes—and adapted to their external environment."[20]

Chapter 2

Place-Based Performance and Ecological Restoration

Mark Dannenhauer

Introduction

Biodiversity and cultural diversity, says V. Shiva, go hand in hand. Shiva writes, "... the cultivation of diversity is the most significant contribution to peace—peace with nature and between diverse peoples. The cultivation of diversity has to be a conscious and creative act...because tolerance alone is not enough to contain the wars unleashed by the intolerance of difference. Diversity is intimately linked to the possibility of self-organization. Decentralization and local democratic control are political corollaries of the the cultivation of diversity. Peace is also derived from conditions in which diverse species and communities have the freedom to self-organize and evolve according to their own need, structures and priorities" (Shiva, 1997).

Ecological restoration is an ongoing human activity comprised of many conscious and creative acts that seek to create the conditions for habitats to self-organize (in contrast to the dominance of invasive species, for example) and evolve according to their own needs, structures, and priorities (as in the successional stages of a chosen reference ecosystem that functions without human assistance).

Place-based performance is a performing art created largely from access to local materials and local experiences resulting from an ongoing dialogue between people and place. It is also an ongoing human activity that seeks to actively cultivate diversity, both biological and cultural. The longevity of place-based performances is like that of forests. Some are old growth, many more are re-growth. Some old growth performing traditions continue today, though many are threatened as old growth forests are threatened. Re-growth performances are also developing today and, from them, old growth forms may yet emerge. The community expression that results from the dialogue between ecological restoration and place-based performance gives new life to the field of performance. In company with much of contemporary art, contemporary performance often seems far-removed from ecological health and integrity. Place-based performance encourages decentralized, public arts, organized by and for the members of the local community. Place-based performances recognizes, indeed depends on, the unique characteristics of local habitats observed over time and the unique characteristics of the local observers interacting with local habitats over time. Place-based performance provides a bridge between people and place, between the artistic and ecological restoration. The presence or absence of placed-based performance can be used to indicate the extent of the success of a restoration project.

Ecological Restoration

Ecological restoration is an ongoing human activity comprised of many conscious and creative acts that seek to create the conditions for habitats to self-organize (in contrast to the dominance of invasive species, for example) and evolve according to their own needs, structures, and priorities (as in the successional stages of a chosen reference ecosystem that functions without human assistance).

The Society for Ecological Restoration Primer on Ecological Restoration offers a recent view of restoration. The primer declares, "Ecological restoration is the process of assisting the recovery of an ecosystem that has been degraded, damaged or destroyed" (SER, 2002). Here, restoration is a human activity that assists ecosystem recovery. Central to the idea of recovery is the idea of a return to a desired ecological trajectory. An ecological trajectory considers both biotic and abiotic factors in a ecosystem and their development over time. Following the establishment of recovery, ecological management takes over from ecological restoration and continues recovery by shaping the ongoing processes of the restored ecosystem. The amount of assistance needed varies in amount and intensity. In the primer's terms, a degraded ecosystem—one that is subtly or gradually changed—may require only little assistance. On the other hand, a damaged ecosystem—one that has been acutely or obviously changed—may require more assistance. A destroyed ecosystem—where all macroscopic life has been removed—requires the most intervention (SER, 2002).

The primer further states, "Ecological restoration is an intentional activity that initiates or accelerates the recovery of an ecosystem with respect to its health, integrity and sustainability" [SER, 2002]. It is essential to note that restoration is a human activity—we humans are the assistants who create the intentional activities that initiate or accelerate recovery. Without human intervention there is no restoration, only the often leisurely flow of ecological succession. A restored ecosystem therefore becomes a cultural ecosystem, one "developed under the joint influences of natural processes and human-imposed organization" (SER, 2002), regardless of its pre-restoration state. Maintenance of the restored site as a sustainable cultural ecosystem—one which is managed to maintain biodiversity and productivity—is not part of ecological restoration but falls to ecological management.

Brown field and gray water restoration deals with ecosystems which range from damaged to destroyed. More intensive restoration and management will likely be needed. Brown field projects have toxic elements and health concerns which are not paramount in restoration projects that move from one natural ecosystem to another. For example, the recent Chicago Wilderness restoration, though controversial, sought to move only from one green ecosystem to another, from current woodland to prior prairie (Gobster, 2000). In contrast, the challenge of the ongoing Chelsea Creek Restoration Project in metropolitan Boston involves a legacy of industrial pollution mixed with a current combination of oil tank farms, road salt and sand piles, exhaust from trains, planes, automobiles, ships and other factors (CCRP, 2005).

The SER primer notes, "some ecosystems, particularly in developing countries, are still managed by traditional, sustainable cultural practices. Reciprocity exists in these cultural ecosystems between cultural activities and ecological processes, such that human actions reinforce ecosystem health and sustainability" (SER, 2002). A culture-to-nature continuum is posited by a wide range of other authors. Sierra Nevadan T. Duane writes from a planner's perspective, "The first theme is of co-evolution and the relationship between Culture and Nature. The two are inseparable, for the landscapes of the modern world have all be touched by human hands" (Duane, 1998). Architectural historian D. Upton writes, "Nature and culture are rhetorical antonyms, but they are nearly always entwined in the landscape" (Upton, 1998). Philosopher-activist-physicist V. Shiva writes, "Biodiversity has been seen as the exclusive domain of conservationists, Yet, nature's diversity converges with cultural diversity. Different cultures have emerged in accordance with different endowments of species in varied ecosystems. They have found diverse ways to conserve and utilize the rich biological wealth of their habitats. New species have been introduced into their ecosystems with careful experimentation and innovation. Biodiversity does not merely symbolize nature's richness; it embodies cultural and intellectual traditions" (Shiva, 1997).

Traditional cultural practices have been shown, though neither necessarily nor universally, to respond to and manage ecosystems and biological diversity. Berkes and Folke estimate that the number of social mechanisms uncovered in their work is "only a fraction of the reservoir of human-environmental adaptations." Yet among this fraction are traditional management practices that include: protection of individual and multiple species and their habitats, restrictions on harvests, maintenance of landscape patchiness, watershed management, accommodation of expected and unexpected disturbance events, accumulation and transmission of relevant knowledge through a variety of cultural institutions and values. Fundamental to these practices is the idea that "[a]ll social and ecological systems have the capacity for self-organization. In linked social-ecological systems, there exists a co-evolutionary relationship between local institutions and the ecosystems in which they are located" (Berkes, 1998). Conservation biologist G. P. Nabhan was presented with two maps with data from United States counties. One showed counties' relative years of residency, the other showed which had the greatest number of threatened or endangered species. "Suddenly," he writes, "I went goggle-eyed: the fit was not perfect, but the correlation between the two patterns was undeniable. Where human populations had stayed in the same place for the greatest duration, fewer plants and animals had become endangered species...In places such as southern California, Florida, southern Nevada, and Hawaii, urbanization and invasion by exotic species have created 'hot spots' of endangered native species" (Nabhan, 1997).

Place-Based Performance

Place-based performance employs a combination of arranged sound, image, movement, and text in live events in public spaces with themes that emerge from a dialogue between local people and place. If you were to attend any of the performances announced in your local newspaper you'd more than likely find them taking place in what is often called a black box. A theatrical black box is a performing space offering control over all physical variables of performance—lighting, seating, staging, geometry and so on. A black box may be a Broadway stage or a community playhouse. Control over performance variables remains highly prized. There are very few natural elements in a black box—no sunlight, no top-soil, no green growing things (if the dressing rooms are occasionally cleaned). Black box theaters are as anthropogenically dominated a landscape as one would ever hope to find. Is it small wonder that performance do not leap to mind when one is casting about for aid with landscape restoration?

Place-based performance is the antithesis of the black box. While some place-based performances do take place in interior spaces, many place-based performances take place outdoors either on or near the source sites they draw on. You can find place-based performance in parades on city streets, in parks and plazas, on grassy river banks, near burning prairie edges. Place-based performance sites are public spaces, open to the air and elements, open to dialogue between people and place. The figures in this article show place-based performance elements in the Australian outback, a street in the United States, a town square in Japan, a public space in Benin, a night time festival stage in Trinidad.

For black box performers, outdoor venues are often viewed as a set of hardships to be overcome. It's too hard to project one's voice, the surroundings sound too loudly, the background is too busy or the wrong color, the weather's too hot or too wet. In short, there's too much distraction everywhere. For a place-based performer all these elements are there to be utilized as character or set, but all are essential parts of the performance interactions. One's voice may in fact be just a small part of the overall life of the space, so your meaning is conveyed with movement, image, instrumental sound as well. The surrounding sounds may be dominant, let them become part of the musical composition. Use the varied background as a set to be explored, draw out the stories living there. Enjoy the cameo appearances of the weather, played as they say in film credits, by itself. When the downpours come, remember that it's not the first time that a cameo has stolen the show.

There is new thinking about space in ecological theory. T.F.H. Allen, for example, writes that an ecological community is best thought of not as a group of species, nor a concrete Gleasonian entity nor an abstract Clementsian concept, nor a collection of species populations, but as a set of interactions of biotic and abiotic elements having periodic occupancy of space. "An ant and a wolf who happen on the same square meter occupy the surrounding space at very different scales, breaking down the notion of multi-species community as a thing in a place ... Heterogeneous spatial and temporal scales inside both plant and animal communities deny the community a particular place. The way out of the dilemma is not to assign the community to any place in particular... the community is the interaction, not the physical presence, of organisms or populations in a place," (Allen, 1998).

If at least some ecological restoration occurs in spaces where humans are a periodic or even constant occupant, place-based performance then becomes part of that periodic occupancy of space. An on-site place-based performance space becomes part of an ecological and cultural community, a community of shared spirit, space, sensation, a cultivation of both biodiversity and cultural diversity. In the same way that a space changes ecologically with the periodic passage of a herd of deer, a space also changes when performance passes through. Place-based performance becomes another periodic occupant of a given space, simultaneously part of both the ecological and social communities.

In a black box performance, characters may, at most, speak about their concern with nature. There are transplanted black box performances, played outdoors, where the players talk about the right relationship of man and nature yet do not acknowledge either the actual species site or the myriad interactions taking place on site. Place-based performance acknowledges and celebrates the elements of the chosen site, so that these elements becomes characters, so that both audience and performer can see, feel, hear the site and their interactions with it. Let us consider two types of place-based performance: old growth performance—place-based performance traditions around the world that have persisted for centuries and re-growth performances that have been developed recently.

Musician B. Krause has recorded in ecosystems around the world. He argues for the idea of biophony, that natural systems have their own distinctive soundprints. Instead of trying to record a particular species, Krause records the entire soundscape of an area. By analyzing digital printouts of the soundscape, Krause argues that a given area has a particular soundprint, consistent from year to year. The creatures in the soundscape have unique combinations of pitch and rhythm such that species have their own sonic, as well as physical, niche. Krause has travelled with the Pitjanjara of central Australia and the Jivaro of South America both of whom navigate by the sounds of locally-distinct sonic landscapes. He also cites the work of L. Sarno with the Bayaka of central Africa. Not only do the Bayaka recognize the elements of local soundscapes, they also slot their own music into the available sonic space so that the music interlocks with the natural sonic system. This interconnection is so pronounced that the Bayakan musicians have difficulty performing when called to tour outside their homeland as a significant part of their musical conception is absent (Krause, 1998).

When habitats are damaged, degraded, or destroyed—the conditions that require ecological restoration—the soundscapes of those habitats are damage, degraded, or destroyed as well. If humans have integrated their music into the local soundscape, that music would be lost as well. Krause presents spectrographs, soundprints, of old-growth performance and re-growth habitats. The spectrographs provide a visual impression of the differences in the two habitats and provide a visual metaphor useful when comparing the richness of old growth and re-growth performance (Krause, 2002).

As old growth forests were long thought to be vertical deserts (Luoma, 1999), old growth performances were long thought of as artistic deserts, or at best curious bits of folklore. As part of a general disenfranchisement of traditional knowledge, both ecological and artistic, old growth performances often were not even called art, but rather relegated to categories such as craft, folklore, ritual, or ceremony. These old growth performances are a form of traditional knowledge and practice which are a bridge between nature and culture. Now, as we discover the abundance of life and relationships in old growth canopies, fallen logs, snags, and mycorrhizosphere (Luoma, 1999), we also discover the depth of life and relationships in old growth performance. As we begin to recognize the value of traditional ecological and artistic knowledge, we enrich our conception of art.

Place-based performance includes four key elements: longevity, local access, community expression, and the bridge:

Longevity

Place-based performance benefits from persisting over time, depends on continued engagement of people with place. Although the movements of masker or puppeteer seem so very concrete during a performance, the overall performance is ephemeral in the extreme. Once completed, it disappears. Only repeated performances over time give performances solidity, vestiges of permanence. When in the mummer's play *Saint George and the Dragon,* Saint George is brought back to life via a magic potion, he no more dormant lies. Moreover, both the play and the play's metaphor that connects biodiversity and cultural diversity live again. Without repetition over generations, performance dies. As performance lives over time, beyond generations, performance becomes a key element of the history of a place, a key element in the creation of community.

Old growth performances, like old growth forests, can be valued simply for their longevity. Consider the Noh theater of Japan. Not only have the texts been preserved since their writing at the time of Shakespeare, but the manner of performing—including the tilt of the head and mask, the slide of the foot, the wave of the fan, the path taken on the stage—have also been preserved. Acting and musical roles were handed down from father to son. Rebecca Teele introduces an interview with a Noh performer, " Beginning with his first performance as a...child player, Kongo Iwao describes his early training, his first use of a mask...and what his father first taught him about wearing a mask. He continues then to explain how at twenty-seven...he was expected to discover the essence of past performances of his father or grandfather" (Teele, 1984). Even though Noh performance is handed down from individual to individual, when the performances are viewed over time they become community expressions first, with individual variations of secondary importance. Most black box theater values individual expression first, whether of director or actor or playwright, with community expression of secondary import.

The Bread and Puppet Theater merely four decades old in contrast to Noh's four plus centuries. I spent a decade with this re-growth theater traversing the undulations of a Vermont hayfield descending to the horseshoe-symmetry of a restored gravel pit. I schlepped giant puppets that appeared and disappeared from spots in the landscape. Some years were spent hidden beneath stinky burlap sacks. Some years I twirled the twelve foot center pole operating a giant washerwoman. I ran flying big white birds in the failing evening light, down over the fields, around the bonfires of empires fallen, out into the dark night. Five years, ten years, each 'the nineteenth annual', then the actual nineteenth annual Domestic Resurrection Circus came to pass. Then more. Over time, performance became ceremony. With more time, more repetition, more habituation, ritual might emerge from ceremony.

Local Access

Old growth performance makes use of the materials at hand. If one lives a mobile existence in sparse environment, those materials can be few. Nevertheless it is possible to create highly expressive imagery using the few materials, paint and plants, at hand. A Native Australian "has a totem-animal or totem-plant; in other words , he is 'related' to a certain animal or plant. The human being is duty-bound to 'increase' the number of his animal or plant 'relatives' by observing certain ceremonies in which incidents of a primeval past are re-enacted. These ceremonies are performed in traditional 'costumes'; that is to say, the bodies of the participants, who represent spirits of old, are painted in ever-varying patterns. (Lommel, 1972).

All old growth is not fixed in time like the Noh, nor as basic as the Australian. The Mancrab is a recent piece, but descended from the old growth tradition of

Carnival in the Caribbean and in Europe. In contrast to the simplicity of the materials and technology in the previous example, the materials and technology employed here are more industrial. This figure is comparable to the newly discovered flora and fauna in old growth forests (Luoma, 1999). Performance also draws on contemporary dialogues with landscape. Bettelheim et al describe a Trinidadian carnival puppet created by Peter Minshall: "slithering in the polluted slime of the island, Mancrab, King Evil himself, represents technology's dark side, its promise of environmental destruction. On each night at the King and Queen competitions of Carnival 1983...this hideous dancing mobile moved to the violent music of East Indian tassa drums. Menacing pincers clawed the air threateningly as the multi-armed monster spun its dance of destruction beneath a canopy of white silk. At the climax of the dance, a compressor pumped red paint up tubes attached to four poles, slowly staining the silk at the corners. Before the dance was done, the pure white silk had turned blood red." Mancrab, an artistic variation on a local species, was finally subdued by King Callaloo, son of Queen Washerwoman and Papa Bois, spirit of the forest. Callaloo is also the national dish of Trinidad. Made from soft-shell crab, coconut meat and milk, curry, hot peppers, and ground leaf of the dasheen plant, the soup is both actual and metaphoric mix of cultural and biological diversity (Nunley, 1988).

The work of the theater company Els Comediants, while not an old growth group, draws heavily on traditional Catalan customs that were outlawed for decades following the Spanish Civil War. Comediants' integrated these customs into contemporary performance and presented this synthesis to large and small Catalan communities. A performance would sometimes be a day-long suite of performances. A small group gets up early, mostly brass and drums. A small but very loud early morning parade wakes once-quiet village streets. Later in the morning, troupe members lay out cans of paint and sheets of cardboard on the town square. An audience gathers; the electric band begins to play. On the spot, the audience builds the scenery and a learns a show telling the well-known story of the driving the Moors out of Spain. At noon, with traditional music, traditional giant puppets, like this sun parade through the traditional streets of the traditional village. Following a break in the heat of the day, the troupe prepares a large stage and larger sound system on the town square for an evening show with masks and rock and roll, giant puppets and a liberal dose of Spike Jones—a dance for everyone blending old growth and new (Comediants, 1988).

Place-based performance is driven by the experiences of the participants, not external texts or global entertainment franchises. In many traditional societies, intensive daily involvement with the natural world was a given. In the course of hunting, fishing, farming, people were constantly in dialogue with local ecosystems. Such dialogue, though increasingly rare, does continue in contemporary life. In the nineteen seventies, for example, traditional Provencal farming life—its rhythms, patterns and processes—was threatened by armies of accountants and politicians and mega-competitors. I spent a summer of Provencal nights, beneath the glistening stars, in the failing heat of a Mediteranean hill town evening, the empty twisting dry dust streets capped by still-warm tiles, the day's eyes and nose lavender-field full, watching Le Theatre de l'Olivier perform *Les Paysans. Les Paysans* presented the travails of a long-time farm family in the south of France to an audience of long-time farm families. But the themes of the performance were only the beginning of the action. The real action came as the audience brought melons, wine and other fruits of the fields, then spent long nights discussing issues and responses with the performers. This scenario was repeated all summer-long in small farming communities throughout Provence and the performances became well-woven into the fabric of the communities and region.

Community Expression

The Matsuri of Japan are another old growth form. "The spectacular Japanese community festivals known as Matsuri are centuries old. Even today, in a society driven by technological advancement, these annual rites continue to function as mechanism for purification and renewal and also to ensure all aspects of communal productivity" (M. C. Berns, *in* Gonick, 2002). Images of cranes and other birds, deer and other animals, flowers, carp, waves and flowing water are all found in matsuri. It is not coincidence that animals and natural forces are included, "...the invocation to the deities and interaction with them occurs. This interaction, which is the most important segment of matsuri, is also the most important rite of Shinto, thought crucial to the regeneration of the people and the land." This view supports the vital interplay of people and place, connected through the arts. Further, however, "It is also believed...that the fertility of the land, as well as the potential of the community, can only be revived by eliminating stale and contaminated forces built up during the recent past." Is this not also the concern of ecological restoration? As simultaneous shared experience, performance offers an alternative method of publishing restoration tenets. Performance offers an invigorating alternative to flyers and newsletters. Restoration goal and objectives, patterns and processes are made visible, simultaneously, to the participating community. The designer E. Tufte decries the contemporary flatland of information composed of paper pages and computer screens. "Escaping this flatland is the essential task of envisioning information—for all the interesting worlds (physical, biological, imaginary, human) that we seek to understand are inevitably and happily multivariate in nature" (Tufte, 1998). Tufte presents a number of design solutions but ignores the obvious. The way to escape flatland is to work in the multiple dimensions of sound, image, and movement. Place-based performance presents Tufte's inevitable and happily multivariate worlds in full dimension. Performance thrives without page or screen, happily and multi-variately transferring information between performer/designer and audience in shared space and time. Performance transfers information in a system simultaneously social and personal. Restoration projects are also shared experiences, in real time and space, far removed from flatlands of paper and computer screen. The more one works with the real-time, real-space of the site, the more possible is a successful solution. The more restorationists and performers work in flatlands the less successful the solution.

Place-based performance invites the presence of a broad range of participants. I have observed the Musketaquid art and nature program of the Emerson Umbrella Center for the Arts in Concord, Massachusetts over the years. It has consistently refused to bow to consultants' advice narrow its organizational focus to a particular niche. Instead the program aspires to work with the broadest spectrum of the human community—children, young adults, grownups, seniors as well as environmental organizations, businesses, religious groups, and social organizations. Musketaquid is also involved with a broad range of local habitats—fields, forest, wetland, riverine, aerial. As amplifier of image and meaning, performance offers the opportunity to play with scale of meaning. One of the biggest puppets made in recent Musketaquid community workshops was a giant caddis larva. An indicator of river water quality, the real-life caddis larva is nearly microscopic. The puppet caddis is fully eight feet tall, green and mean. Most of the Musketaquid puppet critters— whether caddis, frog, dragonfly, or heron—dwarf their human makers. The annual Musketaquid parade includes mammals, birds, amphibians, reptiles, insects, trees, shrubs, flowers, grasses, reeds, seeds. Over the past ten years the puppet parade has grown from three children following a lone pied saxophonist across the Umbrella lawn to an assemblage of at least 800 people, giant puppets, banners, musicians, and other visual and performance elements. The puppet workshop format provides participants with the necessary tools, techniques, and design parameters with which to create and perform with puppets of their own design. Within the limits of the basic framework people make a wide diversity of puppet styles and subjects. Over the years the puppets have grown from single representatives of a species to populations of numerous species. Habitat elements, in the form of banners, flags, and other visual works, have been added. Populations of puppets and habitat elements have been gathered into terrestrial, aquatic, and aerial ecosystems. The assembled ecosystem of puppets and people has become a walking land mosaic, a natural and cultural landscape on parade. Recently, the parade has begun to make connections with restoration activities such as animal crossings for the redevelopment of the very busy nearby highway and the restoration of the Great Meadows, a National Wildlife Refuge.

Very few people watch the parade. In the past few years there have been many more people in the parade than watching the parade. At first, the intent was to create a parade to be seen by the community. Now the intent is to create a parade done by the community. Very brief performances are designed to be played for an audience of the parade participants. With the theme, *Make Way for Wildness,* and an underlying concern with landscape ecology concepts such as fragmentation, perforation, and elimination of habitats, a small group of performers ranged throughout the more than quarter mile long parade fragmenting, perforating and eliminating habitats within the parade. Many of the paraders were then performers and audience simultaneously. The parade ended on the Umbrella lawn by compressing the long line of paraders into a tight inward spiral dance to the accompanying brass band. The Musketaquid Earthday puppet parade is an example of the power of place-based performance that acknowledges old growth forms around the world and creates re-growth forms. Such a connection between nature and culture is not limited to bucolic suburban settings. One workshop participant, a green activist from the neighboring city of Somerville—one of the most densely populated areas in the country—convinced her local arts council to create a parade of urban habitat creatures, to create murals and art images based on local urban habitats in store windows throughout town, and then parade the puppets through the new store window habitats.

The Bridge

Old growth performance has long been a bridge between people and place. Al. F. Roberts writes, "African animal imagery challenges common Western views that culture and nature exist in stark, Manichaean opposition...many African philosophies posit a culture-to-nature continuum, with interlacing instances of nature-within-culture and culture-within-nature. Animals may have souls, be devious, and know magic; they both deserve and require sacred attention from humans who interfere with them....Africans may coexist, integrate, and identify with animals in ways disconcertingly difficult for many Western observers to comprehend, let alone accept...African zoomorphic arts such as masked performances, carved and cast figures, and petroglyphs, often capture this very ambiguity with all its potential" (Roberts, 1995). This attitude persists, writes Roberts, even though there are very few large mammals to be found, save in a few large reserves. Masks, stilts, control rods, and costumes serve to amplify both idea and movement in place-based performance.

Or, consider, for example, an early fall day in Neah Bay on Washington's Olympic Peninsula, a temperate rainforest at the edge of the sea and home to the Makah people. Strangely, it's not raining. Or drizzling. Or misting. Instead, there's late summer sun and giant salmon fillets skewered on three-foot tall sticks smoking juicily next to a heaping pyramid of coals. The car-filled Makah Days parade has ended, the traditional-style canoe race is yet to come. The youngster who had earlier grabbed my eye with his simple cardboard wolf mask has dismounted from the top of the back seat of his parade convertible. Makah dancers and musicians now invite us onlookers to join a public version of a ceremonial wolf dance. We step, step, step, following the serpentine path of song, drum, and lead dancer. Our heads move to and fro, trying to be watchful, looking front then panning right to left, now facing back, flip forward, continue. We don't dance well, but we start to get the idea. Afterwards, I asked Makah sculptor G. Colfax why the wolf was so important to Makah arts. "Because it's the smartest and strongest. I once saw a wolf swim out hunting a seal. All you could see was its nose above the water. It's just the best" (Colfax, 1994). The wolf dance, known well to the eldest dancer and to the youngster in the car parade, is part and parcel of the health and integrity of the local ecosystems.

The bridge can also emerge directly from restoration work. In *Totem Salmon,* F. House describes the decades long evolution of salmon restoration work on the Mattole River in northern California (House, 1999). Two of his long-time collaborators, D. Simpson and J. Lapiner, developed the performance *Queen Salmon* based on their experiences with the salmon restoration. B. Doran tells the story in the *North Coast Journal:*

> "Meanwhile, Simpson was on the river trapping salmon as part of the restoration effort. But his theatrical past hadn't disappeared. "While they were on the traps they used to make up songs, about the hard work, about the salmon," Lapiner said.
>
> Among the early fish-trap songs was one called "My Girlfriend Is a Fisheries Biologist." It became part of Human Nature's first musical com-

edy, Queen Salmon. "We had the Alabama-Pacific forester singing that song," Simpson recalled. "He was in love with the Fish and Game biologist, but she would have nothing to do with him until he started seeing things the right way."

The play debuted in 1991, a time when community conflict was at its height in the Mattole. "People were at each other's throats," as Simpson put it. "We had just endured Redwood Summer; Fish and Game was pushing for a zero-net-sediment discharge. California Department of Forestry coordinated a couple of public meetings. One at the Turf Room in Ferndale drew about 250 people, most of them just there to vent anger — about regulation, about government, about environmentalists. It was really nasty; it nearly turned violent."

"It was all more food for Queen Salmon," Lapiner pointed out. Simpson finished her thought: "We were creating this show about a community in conflict, between the logging and ranching elements and the newcomer environmental restorationists. It showed how a mutual love of salmon brought the community together."

The play's satirical barbs poked fun at both sides; the intent was to get people to laugh at themselves, and through laughter to defuse tension. They took the show on the road with performances at HSU's Van Duzer Theatre; even the PL executives enjoyed it. The question was, did it change anything?

"We like to think it played some role, large or small," Simpson said. After a number of people in the valley had seen the play, a meeting was held at the Mattole Grange. "It was like night and day," Simpson related. "I think people were so appalled by how bad things had gotten that they were ready to sit down and talk to each other."

The show would eventually have three tours up and down the coast with fine-tuning along the way. [Joan Schirle, artistic director of the Dell'Arte International School of Physical Theatre in Blue Lake] marveled at the community effort that went into the traveling productions. "The shows were almost tribal experiences," she said. "It was like they had half the population of Petrolia moving, taking the show on the road. Here you had a group of people connected because they shared the same watershed producing a cultural event. That's not done a lot" (Doran, 2003).

The Lichen Analogy: Performance as Restoration Indicator

The presence/absence of accompanying place-based performance is an indicator of the completeness of ecological restoration. Without place-based performance ecological restoration is incomplete. Place-based restoration not a final decorative step, but an active companion in the restoration process.

Brown field restoration moves from one cultural ecosystem or landscape to another. Restoration without the hand of culture is simply ecological succession. Most sites considered for restoration have been anthropogenically damaged. The principles of physics seem to apply equally to natural or cultural landscapes. The principles of ecology would seem obliged to do the same. As the English ecologist O. L. Gilbert writes, "If the founding fathers of ecology has studied cities rather than the most natural areas available to them they would have given greater prominence to anthropogenic factors [of ecological relationships]" (Gilbert, 1989). Less than one percent of Massachusetts forest is thought to be old-growth forest. The rest of Massachusetts' forested ecosystems are therefore cultural forested ecosystems, a clear result of the co-evolution of nature and culture. D. Foster points out that much of Thoreau's landscape was composed of cultural ecosystems wherein fields and forest edges outnumbered forests in the overall land mosaic. Birders tend to prefer restoration of fields and edges, being better for many bird species. Wolf fanciers tend to prefer restoration of untouched forests, being better for these top carnivores with large range requirements (Foster, 1999).

The idea of historical ecological trajectory is central to the SER primer definition of restoration. Trajectory emphasizes the importance of a restoration project's position in time, the project's relationship to history and to the future. If a historical ecological trajectory may be found, and if nature and culture co-evolve, then there should also be an historical cultural trajectory. Place-based performance, whether annual, seasonal or other periodicity, provides a home for these co-evolved trajectories. Presently, the curves for contemporary place-based performance, in terms of relative frequency and abundance over time, aim sharply downward. Place-based performances are characterized by connection to place over time, by access to local stories, materials and sites, and by community expression. In much of the contemporary United States this combination of qualities is less-often found. Many local newspapers have a weekly section listing local cultural events. Of the total number of cultural activities, how many have persisted over generations? Many towns have community theaters which have strong community membership but how often are the subjects rarely locally-based, built from local materials or responsive to the elements of local ecosystems. In contrast, many towns have nature centers offering interpretive presentations (a type of performance) which are highly site-based but rarely participant-driven. In contrast to community theater, a few to many situation, nature center interpretation is most commonly a presentation by a single expert to a crowd, a one to many situation. The extent to which a restoration project initiates or accelerates the recovery of its cultural, as well as ecological, trajectory is indicated by the extent to which the project initiates or accelerates the presence of place-based performance.

The presence or absence of place-based performance is a useful indicator of the degree of success of restoration projects. The use of performance as a restoration indicator is similar to the use of lichens as indicators of pollution. Lichen biomonitoring offers a cheaper and easier qualitative complement to quantitative methods of atmospheric sampling. "The use of organisms as surrogates for assessing the impact of environmental factors has been widely used in pollution monitoring. Within an ecosystem it is often difficult and expensive to measure the environmental variables and easier to measure the signal from an identified indicator and use this to estimate the environmental condition" (Nimis, 2002).

As the presence or absence of lichens can indicate pollution levels, the presence or absence of place-based performance can indicate the success of restoration projects—the extent to which nature is healed, relationships repaired and to which landscape architecture restores ecological spaces and consciousness. As with lichens, rather than directly monitoring environmental variables in a restoration project, one measures the signal from an identified indicator, in this case—performance, and uses this to estimate the condition of the restoration project. Place-based performance shares many characteristics of lichens. Lichens are ubiquitous, symbiotic and damaged by hurt to either symbiont, perennial, low cost, have a relatively high tolerance for damage, and available to all ages and abilities through different methods (Nimis, 2002). Place-based performance has been found around the world, is a symbiosis of nature and culture and damaged by imbalance of one or the other, is perennial and relatively low cost, has high tolerance for natural and cultural damage, and is also available to all ages and abilities through different methods. Lichens are found in old growth forests and in relatively new growth woods. Lichens don't measure air quality but rather "the effects of pollution and other environmental change on the biotic component of ecosystems" (Nimis, 2002). Similarly, performance monitoring doesn't quantitatively analyze restoration projects, but the effects of restoration projects.

Place-based performance has been found in cultures around the world on all inhabited continents. Place-based performance has been a characteristic of the co-evolution of biological and cultural diversity around the world. Different environmental conditions have provided different raw materials for the arts. Different interactions between cultures, organisms and environments have created different meanings for landscape and performance. Biological diversity plays the role of the photobiont, the magical partner in the symbiosis. Cultural diversity plays the role of the host. If there is damage to the photobiont, if biodiversity is significantly reduced, the performance symbiosis is also diminished. Without a host, biodiversity loses its framework or connection to restoration projects. Rather than restoration, one then has simply cycles of succession and disturbance.

When looking at a restoration, ask if there is place-based performance associated with it. Are there old growth performance elements or re-growth performance elements? What is the age of the performance? Do the performances drawn on local materials, sites, and stories? Are the performances community expressions, characterized by a diversity of points of view and a complex set of relationships, presented in public spaces? Do the performances provide a bridge facilitating travel between natural and social systems, between people and place?

Design by Nature

In *Nature by Design,* E. Higgs advises restorationists to focus on three key goals when creating nature by design: 1) Act to insure the integrity of chosen ecosystems and fidelity to a chosen historical ecosystem condition. 2) Build engagement of people with place. 3) Connect the future with the past. Place-based performance helps restorationists implement these goals. Place-based performance, as a design tool, cultivates ecological and biological diversity and is useful throughout the process of ecological restoration. Place-based performance, as an artistic medium, is informed and shaped by ecological restoration work. Place-based performance becomes design by nature.

Place-based performance cannot insure the integrity of an ecosystem or the fidelity to a chosen historical condition. But place-based performance can emerge from activities that do. Recall the songs created by Mattole River restorationists that led to the larger performance, Queen Salmon. Performance can be a way to spread information of restoration issues, as the themes of Musketaquid parades have tried to do. Performance can be used not only as a culminating celebration but also as a way of working throughout the restoration process. Place-based performance can put restoration issues squarely in the public's eye, as Queen Salmon landed in the midst of intense and prolonged community struggle or the nearly ephemeral parade of local puppet critters jamming Concord's Main Street traffic for 20 minutes of each year.

Place-based performance depends on the engagement of people with place. Place-based performance can be a community expression or a set of individual expressions. But the presence of place-based performance shows that the full range of a community's capacities are engaged in restoration work, restoration of habitats, restoration of human community, and restoration of the arts that long have connected them. Different people have different interests, different skills; restoration benefits from a broad based of activities and perspectives. Place-based performance invites participants to use their artistic skills and sensibilities. Restoration consists not only of histories and scientific studies and political maneuverings but also of performance and the other arts. Place-based performance invites participants to use their artistic as well as historical, scientific, and political skills. The challenge of solving the many creative problems necessary to creating any performance can foster team building among a group of restorationists. Place-based performance opens the door to collaboration between organizations within a community. The Musketaquid Program this year is partnered by Mass Audubon's Drumlin Farm Wildlife Sanctuary, the Organization for the Assabet River, and the Sudbury Valley Trustees. Over the years, Musketaquid has worked with elementary, middle, and high schools, Restore the North Woods, the Great Meadows wildlife refuge, Walden Woods, the local branch of Keeping Track, the town Conservation Committee, and many other community groups and organizations. For this year's Earthday Event, there are opportunities for visual artists in many media, singers, journal keepers, and performers.

Place-based performance is always a dialogue between future, past, and present. Some old growth performances, like Noh, have been preserved almost intact for over five hundred years. Other old growth performances, like Caribbean carnival, has also grown over hundreds of years, but with varying forms that nevertheless are consistent with the overall performance trajectory. Le Theatre de l'Olivier provide local farmers with a situation in which to gather together to talk about the future of their farms, their landscape, their lives. Els Comediants combined traditional performance elements with contemporary performance elements, provided a public situation for community members to build a community expression about a central historical event and connect that event to the present day.

"To design," writes Higgs, "is to work something out in a skillful or artistic way...The notion of design can be pushed further... toward wild design—

that is, the deep appreciation of what an ecosystem requires to flourish, and then making such conditions possible." This last is a crucial point, that design is not always about the making of things or objects but can also be about creating conditions or situations. Design critic R. Caplan claims that the most elegant design of the nineteen fifties was not a car or a chair but the sit-in. "Achieved with a stunning economy of means, and a complete understanding of the function intended and the resources available, it is a form beautifully suited to its urgent task" (Caplan, 2005). Ecological restoration is a way to actively design and cultivate both biological and cultural diversity, to encourage both ecological self-organization and cultural self-determination. Place-based performance is a design tool that helps facilitate the process of ecological restoration, that encourages personal and community engagement with both arts and ecosystems, precipitates and results from action focused on ecological restoration, and mediates between past and future along both ecological and artistic trajectories. Perhaps the symbiosis of ecological restoration and place-based performance will be regarded by future generations as having been the most elegant design solutions of the early twenty first century.

Chapter 3

As Inside, So Outside: Restoration of Inner and Outer Landscapes

Lane K. Conn and Sarah A. Conn

"We need a new way to envision our relationship to the full panorama of the crawling, burrowing, swimming, gliding, flying, circulating, flowing, rooted and embedded Earth. We need to be and to feel differently, as well as to think and believe differently" (DeQuincey, 2002).

The restoration of ecological consciousness "inside" humans is a crucial aspect of any work toward restoration of the health of the "outside" nonhuman natural world. The two - inner and outer restoration - are inextricably intertwined and affect each other fundamentally. Even the use of the phrase "inner and outer" masks the fact that these two realms are part of the same fundamental reality.

Recognizing and working with the connection between the landscapes of the human psyche and the landscapes of the more-than-human natural world in ways that articulate the health of both is the primary work of ecopsychology. This discipline brings ecology and psychology together to develop new understandings of health and society which emerge from attending to the human place within the earth as a living system.

Ecology is the study of connection, of the interrelationships among all forms of life and the physical environment. A basic subject of ecology is the "ecosystem," defined in the 1950's by E. Odum, considered a father of ecology, as "any unit that includes all of the organisms (i.e., the 'community') in a given area interacting with the physical environment" in ways that maintain the biodiversity necessary for adequate flows of nourishment in the community (Odum, 1971).

Psychology is the study of human experience - perceptions, feelings, thoughts, images and actions, as well as intuition, spirituality and mindfulness - as the human individual makes contact within intimate and social systems. Ecopsychology brings psychology and ecology together to expand the range of both disciplines in studying the human psyche or soul within larger natural systems. The task of ecopsychology is to look at the interconnectedness and interdependence among psychological, cultural and nonhuman natural systems. "Ecopsychology is rooted in the nonduality of humans and nature. Human consciousness and the human psyche are expressions of natural processes along with the rest of nature" (Davis, 2000).

How is the ecology of human experience—all the perceptions, feelings, thoughts, images, intuitions and actions within and among humans in intimate and social systems—affected by and affecting of the sustainable and mutually-enhancing flows of nourishment in the natural ecosystems of which humans are a part? As P. Shephard, human ecologist, suggested in 1973, the environmental crisis can be seen as signifying "a crippled state of consciousness". How have we become trapped in mental constructs and limited ways of knowing that degrade both our consciousness and the social and natural processes within which we live? In order to attempt an answer to these questions, we need to attend to the current state of human consciousness. Then we will explore other ways of knowing and connecting with the greater intelligence around us in the earth as a living system, in order to contribute to the restoration of both inner and outer landscapes.

Ecopsychology is a new perspective on phenomena that traditional ecology and psychology have either incompletely addressed, completely neglected, or relegated to the realm of epiphenomena. This new perspective is based on ways of knowing that neither mainstream ecology nor mainstream psychology have addressed. Ecopsychology is concerned with introspection as well as inspection, contemplation as well as experimentation, quality as well as quantity. Cognitive understanding is augmented by intuition, resonance, and imagination, by spiritual as well as sensory ways of knowing. The interior, in-between and relational aspects of the world are the focus of attention as well as the exterior, discreet parts. Ecopsychology's approach to phenomena is about opening to their manifestation and resonating with them, rather than making them calculable and taking their measure.

Ecological health is the larger context of human health; they go together as outside-inside. Healthy human functioning in an ecopsychological context includes sustainable and mutually-enhancing relations not just at the intrapersonal level (within humans) or the interpersonal level (among humans) but also at the level of "interbeing" (Nhat Hanh, 1987) (between humans and the non-human world). The contextual background for healthy human functioning which supports sustainable relations is an ecological mode of consciousness.

> "The deepest cause of the present devastation is found in a mode of consciousness that has established a radical discontinuity between the human and other modes of being and the bestowal of all rights on the human" (Berry, 1999)

Consciousness has been defined in a variety of ways, and is a field in and of itself. We are using it here to refer to awareness of existence, sensations and surroundings in a moment-to-moment integration of experience in the present. Modern consciousness in our industrial culture tends to be dominated by the analytical mode, restricted to linear, sequential, mechanical ways of knowing and of experiencing the present. This mode of consciousness tends to separate humans from nature and to elevate reason above all other ways of interacting with the world. This bias towards separation and reason has been with us for many centuries. The Eleusinian Mysteries in ancient Greece spoke against the split between humans and nature by using the phrase, 'As above, so below' (Bly, 1980). Centuries later, Einstein (*in* Wilber, 1981) referred to this split as an "optical delusion of our consciousness."

To recover from this optical delusion, we need to restore an ecological mode of consciousness, one which includes ways of knowing which are holistic, nonlinear, dynamic and intuitive. Ecological consciousness locates the human psyche or soul within the larger context. At its center is the "...intuitive awareness of the oneness of all life, the interdependence of its multiple manifestations, its cycles of change and transformation" (Capra, 1988). In this context, consciousness is not the exclusive possession of the human being. Again, from the Eleusinian Mysteries: "The Mysteries evoked an awe toward matter, a sense that we shared a consciousness with plants, animals, and stones, and that all of these shared a consciousness with the 'soul of the world' ...." (Bly, 1980).

Within ecopsychology, two interconnected areas of healthy human functioning which support ecological consciousness are as follows: 1) diversity in modes of contact and ways of knowing, and 2) embeddedness in community with all living beings and the land. The goal of ecopsychological practice is to develop methods and forms which enable individuals to sense, think, feel, imagine and act as interdependent beings, interconnected within the whole community of life and land at all levels.

The restoration of ecological consciousness requires attention first to our modes of contact and ways of knowing. We are beginning as a species to recognize the constriction in the modern consciousness and the effects it has wrought. The modern industrial paradigm encourages us to view nature primarily as a commodity for consumption or manipulation. When we are unable to dominate nature, we view even that in anthropocentric terms, such as the weatherman who announces that "The snow is falling out of control in Vermont!" (Kidner, 2001). Organic wholes are broken into their component material parts. All things, including humans, are individualized and defined in commodity terms (Taussig, 1980). What effect does this dominant mode of consciousness have on our experience of the nonhuman world? What do we see, feel and think when we gaze upon a beautiful "natural" landscape? In a network documentary made in 1990, a leading Northwest timber executive describes looking at trees as "stacks of money standing on stumps". What he sees when looking at these natural beings are objects for human use. And what is our experience when we perceive a degraded landscape? Do we automatically begin planning to "fix" it, to make it pretty to us, accessible for our use?

Nature as viewed through the lens of industrial/objectivist consciousness is little more than a filling station, a super shopping mall for humans. An example of this is the agreement signed by the Department of the Interior allowing the Diversa Company, a California biotechnology firm, to "remove certain wildlife from Yellowstone Park and patent it to use for commercial purposes" (Kimbrell, 2002). When a suit was brought against them, the court accepted Diversa's definitions: a) " that Yellowstone is an 'outdoor laboratory' and b) that the wildlife being transferred is patentable 'technology'". Thus, the mechanistic view of life becomes encoded not just into our consciousness but into our legal statutes.

> "Mr. Chairman, if you know wilderness in the way you know love, you would be unwilling to let it go. We are talking about the body of the beloved, not real estate." (T. T. Williams, 2002: 1995 testimony before Congress)

What is real? How do we know? The objectivist paradigm of modern science holds that there is a reality "out there," and that there is a separation between the observer of nature and nature itself. Objectivist science claims that it is possible for the observer to know reality if specific rules for defining evidence and particular methodologies for exploring the world are followed. With these lenses, we are likely to view nature, whether beautiful or degraded, as a reality separate from us, a thing "out there." Our ways of knowing within this context are to measure, to manipulate and to fix in order to use for our benefit. We can know nature, but only as radically discontinuous from humans, devoid of being or consciousness, "despiritualized and depersonified" (Fisher, 2002).

The postmodern deconstructionist paradigm challenges the lens of the objectivist paradigm underlying this dominant modern consciousness. Although like the objectivists they posit a schism between humans and the rest of nature, the deconstructionists maintain that the objectivist claim is merely one of many narratives competing for authority about reality. In this view, we humans can never know reality, or even if it exists. Our representations of the world are merely interpretations; our experience is determined by social and cultural filters which shape and distort our perceptions. In this view, there is no "nature" separate from our texts and narratives about it. All attempts to talk about nature are therefore merely competition for authority (Soule and Lease, 1995).

The deconstructionist paradigm does a service by creating an opening for more varied methodologies and diverse ways of knowing than have been endorsed by objectivist consciousness of the modern industrial world. A major problem with the deconstructionist way of thinking, however, is that it dismisses nature as mere narrative, a social construction of humans. Our experience of communing with nature, all our sensual delights in nature, our pain and concern about a degraded landscape - all are but cognitive constructions, not so different from delusional hallucinations. This may qualify as the ultimate in anthropocentrism, a clear-cutting of consciousness in the world. As Shephard (1995) suggests, the deconstructionists suffer from an atrophy of the senses, cutting us off even further from the earth as a living system.

An anthropocentric view of the natural world, whether objectivist or deconstructionist, results in degradation of both outer and inner landscapes. In the words of Swimme and Berry (1992): "What humans do to the outer world, they do to their own interior world. As the natural world recedes in its diversity and abundance, so the human finds itself impoverished in its economic resources, in its imaginative powers, in its human sensitivities, and in significant aspects of its intellectual intuitions". As the diversity and complexity of the "outer" world decreases, so does the flexibility and creativity of the "inner" world of human experience. And vice versa: as we become more entrained into industrial consciousness, we contribute to decreasing the diversity and complexity of the nonhuman world. Thus are inner and outer landscape restoration fundamentally linked.

> "Our behavior is a function of our experience. We act according to the way we see things. If our experience is destroyed, our behavior will be destroyed. If our experience is destroyed, we have lost our own selves." (Laing, 1967)

Ecopsychology agrees with the objectivist viewpoint that nature exists beyond the narratives and perceptions of humans and with the deconstructionists that our perceptions are affected by our constructs. However, the challenge is in recognizing and diversifying our ways of "knowing" nature. Through the development of ecological consciousness, we can expand our knowing of nature and connect with it in ways other than those sanctioned by mainstream modes of thinking. This does not mean the abandonment of mechanistic thinking, but rather seeing it as one style of thought among many ways of accessing what's so.

As the deconstructionists state, what we perceive will be shaped by the questions we ask, the assumptions underlying them and the methods used for answering them. Huxley (1954) noted that the main function of the human brain and nervous system in the role of perception is to eliminate input in order to keep us from being overwhelmed. "What comes out the other end is a measly trickle of the kind of consciousness that will help us stay alive on ... this planet. To formulate and express this reduced awareness, man has invented and endlessly elaborated upon those symbol-systems and implicit philosophies which we call languages...[This] confirms in him the belief that reduced awareness is the only awareness...so that he is all too apt to take his concepts for data, his words for actual things".

We must move out of the assumption that humans are separate from nature, that "inside" and "outside" are two different realms. This means recognizing "the human-nature relationship as a relationship...granting the natural world psychological status; regarding other-than-human beings as true interactants in life" (Fisher, 2002). The challenge is finding ways of recognizing this relationship, of going beyond the usual assumptions by restoring non-conceptually mediated methods of direct knowing. These other ways of knowing are essential to the restoration of healthy ecological consciousness and to the restoration of nonhuman natural landscapes.

What do we mean by direct ways of knowing which enable ecological consciousness? We are talking about direct experiences not mediated by cognition, interpretation or projection. For example, consider the calming and soothing direct experience of gently falling rain, or the startle response triggered by thunder and lightening. For these to take place does not require past experience. Here there is an isomorphism in the relationship between physical expression "out there" and experience "inside" the observer. Isomorphism refers to the possibility that "processes which take place in different media may be nevertheless similar in their structural organization" (Arheim, 1961). Thus there is a direct connection between physical forces in the world and the dynamics of consciousness in the human.

Another way of understanding this direct way of knowing is through the distinction Heidegger (1971) makes between language as disclosure and language as representation. In order not to restrict our consciousness by taking our concepts as direct data, as Huxley described above, language as disclosure is a primary way of communicating. "Saying" what one perceives, in this mode, is bringing something to light, disclosing it, uncovering it from within the experience of it in the moment. It is not representing or standing for something, not a linguistic expression added onto the phenomenon after it appeared. Language as disclosure arises out of the direct experience of encountering the phenomenon.

What happens when we "perceive" a "tree" in the forest? We are saying here that perception is something we and the tree create together through an active interchange. The tree is not simply out there in the forest coming to us through the passive receptors of our senses, as the mainstream scientific view holds. Nor is it simply inside us, created by our expectations and our texts, as the deconstructionists maintain. The color of the red rose is neither "out there" in the rose, nor "inside" the eye of the beholder. This "either-or" way of thinking is what A. N. Whitehead (1955) calls the "fallacy of simple location". Instead, perception is a relational event, a process that happens neither inside nor outside but between us and the beings around us (Whitehead, 1955). If we are to restore our ecological consciousness, we must develop a discipline and practice which enables us to experience a mutual relationship with nature. Only in a such a relational interaction can we hope to make conscious the experience of interconnection in a way that will function in our behavior.

> "As the cricket's soft autumn hum is to us, so are we to the trees as are they to the rocks and the hills." (Snyder, 1992)

Each being, human or nonhuman, has a display, an expression of presence in the world which emanates from its particular structural organization in a particular moment. Gestalt psychology refers to this as the "expression that is perceptually self-evident," existing apart from psychological processes such as learning, past experience, motive, interpretation, association, projection (Arheim, 1961). The world around us is thus active and alive, inviting us to notice. A natural being presents itself to us through its display, which calls for our attention. We can relate directly to it by resonating with it. If we are open in the moment to its call, then we are able to engage directly with it. Our everyday language expresses this when we say, "Oh, that tree (or mountain, house, stream, leaf, bridge, rock ...) caught my attention!" When our attention is captured by the display of the other, we and the other can enter into a relationship, a kind of conversation.

As we discussed earlier, the active, assertive, analytical mode of consciousness, dominant in our culture and supported by our language, teaches us how to manipulate physical bodies, perceive boundaries, analyze and divide the world into separate and distinct objects. This mode of consciousness tends towards the verbal, analytical, sequential and logical, and supports the discontinuity between humans and the nonhuman natural world. It tends not to recognize that humans have ways of directly experiencing a nature that is outside of our constructions of it through the ecological mode of consciousness, which is eco-sensitive and receptive rather than ego-assertive and active.

The ecological mode of consciousness allows for direct contact with the other, for us to be open to its expression or display without preconceived notions of what will show up. "When you understand all about the sun and all about the atmosphere and all about the rotation of the earth, you may still miss the radiance of the sunset. There is no substitute for the direct perception of the concrete achievement of a thing in its actuality" (Whitehead *in* Watts, 1991). The receptive mode of consciousness tends towards the nonverbal, holistic, nonlinear, intuitive. The stance is one of opening to the other, taking in whatever is presented, dwelling in the phenomenon. This is the natural way that most children experience the world. In this mode of consciousness, there is no dualism, no sharp division between inner and outer, between animate and inanimate, between conscious or non-conscious, between body and mind, between public and private, between human and more-than-human nature. This unbroken continuity between self and nature is an enchanted world, filled with energy that some cultures call "spirits."

This experience of unity with nature is not so much rare as it is unreported because it does not fit the dominant paradigm. For example, the scientific world was stunned when B. McClintock, who won the Nobel Prize in 1983 for her research on corn plant chromosomes, described her approach as becoming intimately connected to each corn plant by listening to its story (Jensen, 2002). Her study of the corn plant chromosomes, so small that no one else had been able to identify them, involved experiencing them getting larger and larger the more she worked with them. "I wasn't outside, I was down there - I was part of the system....I actually felt as if I were down there and these were my friends" (E. Keller, in Winter, 1996).

Another example of basic knowledge gleaned from the natural world through this direct way of knowing is reported by Jeremy Narby. As a participant observer, Narby (1998) studied the shamans in the Peruvian Amazon's Pichis Valley. He found that these shamans, without any exposure to Western scientific ways of investigating and conceptualizing, had been able to accumulate vast knowledge of the healing properties of plants. How do these shamans acquire such knowledge? Narby found that through a process that included fasting and drinking a mixture of plant juices called ayahusca, the shamans communicated directly with the plants in a state of defocalized consciousness. "In their visions shamans manage to take their consciousness down to the molecular level".

The shamans' direct communication with the nonhuman world can be seen as a direct experience of shared consciousness between humans and nature. The visual representation of what the shamans call "yoshi," the spirit or animate essence of all life, human and nonhuman, bears a striking resemblance to the visual representation of DNA, or the double helix from Western science. What the shamans have experienced is later replicated by Western science with entirely different methods: "There was indeed DNA inside the human brain, as well as in the outside world of plants, given that the molecule of life containing genetic information is the same for all species. DNA could thus be considered a source of information that is both external and internal...." (Narby, 1998).

In mainstream scientific thought, "knowing" through shared consciousness between humans and nature is not possible, because modern science "is founded on the notion that nature is not animated by an intelligence and therefore cannot communicate" (Narby, 1998). "On the contrary, this sort of direct experience recognizes a world that is fully personalized - full of spirit, feeling, intelligence, relation; a world that in its diversity and fascination shames the tendency of the human ego to categorize or explain" (Kidner, 2001) Subjectivity is much larger than we usually assume, present in ecological systems at all levels (Bateson,1972).

Another way of knowing directly, or opening to the subjectivity or aliveness of the other, whether human or non-human, is through intuition. Intuition is "knowledge without recourse to inference" (Ornstein, 1983), knowledge through direct, immediate apprehension. "[W]hereas the verbal-intellectual mind withdraws from the sensory aspect of the phenomenon into abstraction and generality, the intuitive mind goes into and through the sensory surface of the phenomenon to perceive it in its own depth" (Bortoft, 1996). This form of "perception" is a kind of resonance, a way of knowing we will be describing more fully below.

When we open to the other - human, non-human natural being, or landscape - through these direct ways of knowing, who are we? Gestalt therapy views the self as the process of experiencing the present moment, in which we are "inextricably caught in a web of relationship with all things," (Latner, 1992). The emphasis in ecology on diversity, complexity and flows of nourishment within the organism-environment field (which includes non-human beings and the land) can enhance our understanding of what constitutes a healthy "web of relationship." The biosphere as a whole, for example, can be considered as "an animate self-sustaining entity," "an intertwined, and actively intertwining, lattice of mutually dependent phenomena, both sensorial and sentient, of which our own sensing bodies are a part" (Abram, 1996). Abram proposes that everything we see, hear, smell, taste, and touch is informing our bodies of the internal state of that "other," that vaster physiology of the living earth, our larger bodies, our "whole-earth part" (Armstrong, 1995). We are not separate from nature; we are simply distinct. We are one whole-part in a universe of other whole-part beings. "We are not alone - we are not uniquely special" (deQuincey, 2002).

Ecopsychology invites us to include all aspects of this web of relationship as important aspects of our larger selves, as the community in which we are embedded. As we mentioned earlier, an ecopsychological perspective on health includes experiencing our embeddedness in community, both human and non-human. In this view, communities exist at every level, and our direct ways of knowing can enable us to experience ourselves as connected within each level. A. Koestler (1978) coined the word "holon" to refer to the fact that systems at all levels—whether cell, organ, human body, ecosystem, biosphere, universe—are simultaneously both wholes and parts. Thus, all living systems, from cell to human to planet and beyond, act as autonomous self governing wholes containing sub-parts and at the same time as dependent parts of larger wholes. Holons manifest both self-asserting tendencies—as a self-contained, unique wholes—

and integrating tendencies—as dependent parts of larger wholes.

Measurements of health and success in most current fields of endeavor have largely been a matter of ego-assertive, differentiating (whole) functioning. Eco-sensitive, integrating (part) functioning has tended to receive only passing notice. A holonic perspective allows for a way of talking about individualism and interconnectedness that does not set up a duality of structure and function. All living organisms are interconnected and interdependent, with both self-assertive and participatory tendencies. This perspective allows for distinction without separation: each part retains its clarity and eminence while remaining connected with and embedded in the larger whole (Conn and Conn, 1998). As A. Watts (1972) has pointed out, the wave is not separate from the ocean merely because it is distinct from it.

One way to apply this holonic perspective to our understanding of the task of inner and outer landscape restoration, then, is to view our experiences in relation to the nonhuman landscape as our distinct, unique ways of being part of those larger wholes. A dominant consciousness that has under-emphasized the part aspects of knowing needs to move towards a stress on the participatory functions, such as curiosity, resonance, intuition, awe, wonder and imagination in relationship to the non-human landscape. The restoration of ecological consciousness requires reactivating diverse ways of knowing so that we can broaden and deepen the connection to our landscapes.

When we open to the other from this holonic sense of self, from this ecological consciousness, our experience may then include a wide, field-like sense of self which ultimately includes all life-forms, ecosystems, the earth as a whole and beyond to the cosmos. We are then able to resonate with other life-forms who have value in their own right, not because they are useful to humans but because they are a part of humans. All beings and larger systems are distinct in their own right and are seen as part of one's larger self, as everything is interconnected and interdependent. Understood in this way, caring for other beings is not an altruistic or selfless act; it is a form of self-care. It is a manifestation of ecological consciousness.

> "We can't control systems or figure them out. But we can dance with them!" (Meadows, 2001)

We have argued that nature is real, existing outside of our constructions and projections, and that we can know it in direct ways. Now let us turn to the question of restoration, defined by Webster's Dictionary as the act of bringing something back to a previous condition or to an unimpaired or improved condition. Major questions arise from this definition: who determines to what previous condition nature is to be returned, and what criteria are to be used to determine what is an unimpaired or improved condition of a natural landscape? If we are not fully mindful of our inner responses to the landscape, if we are not open to the range of connections offered by it, our relationship to it will very likely be determined by the constricted modern industrial consciousness we have all been entrained to develop. Restoration which proceeds from dominant, anthropocentric ways of knowing will produce landscapes that are projections onto rather than perceptions of nature, reinforcing the notion that humans are apart from rather than a part of nature.

If restoration is an attempt within this context to "put back" what once was, it lacks a vision born out of a relationship with nature. A mutual relationship, based on direct knowing of the landscape as it is, including the ecological and cultural history it holds, would engender an understanding that nature does not merely replace, it elaborates and continuously alters the landscape. Nature is not efficient but abundant, not contained but overflowing, continuously evolving.

> "The future can't be predicted, but it can be envisioned and brought lovingly into being. Systems can't be controlled, but they can be designed and redesigned" (Meadows, 2001)

In order to restore a landscape, we need ways to design and redesign, to "dance" with the system. As we have been proposing, in order to be in a mutual, flowing relationship with a landscape, we must develop ways of knowing that connect us intimately with it as it is. For example, within this context, "thinking" would "not refer exclusively to deductive reasoning, calculating, or categorizing. These dualistic operations reduce beings to objects for the subject. In genuine thinking, the 'self' or 'subject' disappears" (Zimmerman, 1981). Thinking as part of ecological consciousness is a kind of "thanking" (Heidegger, 1968), celebrating the gift of existence, ours and all beings. When we engage in this kind of thinking, we are opening our hearts, as well as our minds and senses, to other beings. We can then experience the world as a basic interrelated and interdependent unity, an experience of "oneness" that invites spiritual aliveness.

A way of knowing that informs intuition and enables a mutual relationship with the other is described by Kidner (2001) as "resonance". As we have noted above, resonance is a way of connecting directly with the structural flow of another being or landscape. Resonance is a felt way of experiencing our embeddedness in the more-than-human as well as the human community, including awareness of being simultaneously a whole and a part of a larger whole. Resonance refers to a way of being in the world by releasing subjectivity to all beings. We are then able to relate as subject to subject, mutually, rather than as subject to object. This involves not only a shift in epistemology (how we know) but also in ontology (what is to be known). Central to resonance knowing is the recognition of and respect for the other as having a standing in its own right. Resonance is about interaction and integration, about being open to and joining in the dance with the other.

How might the restoration of ecological consciousness impact the task of landscape restoration? If landscape architects come to a site with this consciousness, they will be able to "consult the site," resonating with the beings and their connections in a way that enables a true mutual partnership, a community endeavor that includes the nonhuman in the process. One way to know the whole of the landscape is by going into the parts fully to intuit or resonate with the relationships that make the whole, not just distancing oneself to get an intellectual overview. To know nature, one must step into the parts, enter into the nesting of the whole. This involves intentional consciousness without assertive or willed intention. It enables opening into a reciprocal, participatory, communal relationship with the site.

One process that can enable this way of relating is a process of direct knowing developed by Conn (2002) called "Opening to the Other," followed by a "Council of All Beings" developed by J. Seed and J. Macy (1988). "Opening to the Other" is a process which invites us to take a step outside our usual and customary ways of being conscious.

> "Stand still. The trees ahead and the bushes beside you Are not lost. Wherever you are is called Here, And you must treat it as a powerful stranger, Must ask permission to know it and be known...." (Brown, 1994)

To open to the other fully, we are invited to put aside our present notions and habits for relating to non-human beings. We begin by slowing down, leaning back from our usual way of being with the world around us, to take what Heidegger has called a "step-back-before the arrival of the other beings" (1972). This requires finding ways to enable ourselves to be open and receptive to the other, freeing its display, its approach, its arrival. Being curious and willing to have experiences that go beyond the familiar invites us to adopt a beginners mind, to give our mind a "coffee break" so that we may come to our senses and to other ways of knowing.

What we are opening ourselves to is the direct experience of the other, requiring that we shift our sense awareness from a precipitating to a participating mode of interaction, from making it happen to sharing in the happening. It will nurture our mindfulness of the other if we switch from looking to seeing, from listening to hearing, from touching to being touched.

Usually we restrict our sense awareness through our intentions. We look for something, we listen for something, we reach out to touch something. Essentially we are shaping the meeting with the other. But suppose we see instead of looking, hear instead of listening, and are touched in place of touching?

We can then experience the breeze touching our faces. We can notice how it brushes against our skin, making its presence known to us directly. We can experience our contact with the ground through our feet directly by reversing the assumption that we are touching the ground with the soles of our feet, instead experiencing the ground coming up to meet us. And we can turn our gaze toward leaf or blade of grass and allow it to present itself to us, to imagine it coming over and into our consciousness instead of our going out and getting it. We allow other beings to knock on the doors of our awareness, to visit on their own terms and in their own language.

To have the other as a guest in our house of consciousness requires our attending to the other's expression of itself. To do this we have to let go of the words and stories we have for the other. These symbols are no more the other than is the map the territory or the menu the food. Instead of throwing out a net of words to capture the other, we can open ourselves and let the other reveal itself to us.

When we open ourselves in this way, it not only opens us to the display of the other, it also opens our display to the other, whether human and nonhuman. One example of this occurred during a course on "Sustainable Design as a Way of Thinking" which we taught in downtown Boston. During one class, we sent students out to experience opening to the other as described above. One woman experienced opening to a particular tree, spending the time with it as it presented itself to her. As she was returning to class, she was approached by an eight year old girl who said she had been lost for a long time. The girl said she had been afraid to ask anyone for help until she saw this women walking along the street slowly. The girl did not know her own address, just that she lived with her grandmother next to a church in Dorchester, some distance away from downtown Boston. The student took her to a nearby fire station to ask for help. The people at the fire station were able to find the grandmother because she had put out a missing persons report, and the girl was reunited with her home. Through her experience of connecting with the display of the non-human natural world, this student slowed down to what may have been a more "natural" rhythm, thus "displaying" herself to the lost girl as an open person who was safe to approach.

Our embeddedness in the modern technological consciousness has greatly increased the speed of our modes of perception and our ways of being. We become hyperactive, or as Conn (1996) has suggested, part of manic episodes occuring in an entire geographic region. Opening to the other in the way we are proposing requires a marked change in rhythm, a slowing down to meet the natural rhythms in the more-than-human landscape around us. With these guidelines for "opening to the other," we can enter into a relationship with a particular landscape that enables us to give it our full presence, heal our separation from it, and know our interconnectedness directly.

In this ecological consciousness, one way of "consulting" a particular landscape or site is a process developed by John Seed which asks us to, in this state of openness, be "chosen" by a natural being within a site and then to speak as that being in a "Council of All Beings" (Seed et al, 1988; Macy and Brown, 1998). In the case of the student above, she spoke in the class Council of All Beings as a "lost girl," and joined the others who were speaking as trees, grasses, rocks, squirrels and birds as they imaginatively "informed" humans of their perspective on that particular city landscape.

This practice can be useful for architectural design and landscape restoration, as demonstrated in the following example.. A group of architects were planning for a new park building in Tilden Regional Park near Berkeley, California (Lane, 1998) The participants spoke in Council as the beings from the site who had chosen them, advising architects from the perspective of ant, stone, redwood tree, sun, groundhog, and others. In this method, the "client" expands to include the community of non-human natural beings in the site. "Specifically, the Council of All Beings facilitates listening and a process of knowing ecological networks and native biota. These feedback loops of knowledge inform sustainable design, limitations of human settlement, and, where applicable, an architecture that is living, dynamic, and congruent with ecology".

Restoration must be part of an ecological design revolution which includes ecological consciousness, manifested in direct ways of experiencing our embeddedness in all levels of community. Ecological design has been described as "any form of design that minimizes environmentally destructive impacts by integrating itself with living processes" (Van der Ryn and Cowan, 1995) and as "the careful meshing of human purposes with the larger patterns and flows of the natural world and the study of those patterns and flows to inform human actions" (Orr, 2002).

"The standard for ecological design is neither efficiency nor productivity but health, beginning with that of the soil and extending upward through plants, animals, and people. It is impossible to impair health at any level without affecting it all other levels. The etymology of the word 'health' reveals its connection to other words such as healing, wholeness, and holy. Ecological design is an art by which we aim to restore and maintain the wholeness of the entire fabric of life....." (Orr, 2002). This "art" requires the restoration of ecological consciousness in humans so that we recognize ourselves as integral parts of the fabric of life and recognize other beings as parts of ourselves. Ultimately, restoration is about restoring connections among the parts, and ecological consciousness recognizes that "the parts are healthy insofar as they are joined harmoniously to the whole" (Wendell Berry, quoted in Flemons, 1991). With this consciousness restored, we can then participate with the non-human natural world in the task of restoring the landscapes that are part of our larger selves.

> "..................Stand still. .The forest knows Where you are. You must let it find you." (Wagoner *in* Brown, 1994)

Chapter 4

Nature's Memory: Restoration and The Triumph of the Cognitive

David W. Kidner

"The theory which I am urging admits a greater ultimate mystery and a deeper ignorance. The past and the future meet and mingle in the ill-defined present. The passage of nature ... has no narrow ledge of definite instantaneous present within which to operate. Its operative presence which is now urging nature forward must be sought for throughout the whole, in the remotest past as well as in the narrowest breadth of any present duration. Perhaps also in the unrealised future. Perhaps also in the future which might be as well as the actual future which will be." (A. N. Whitehead)

The enrichment of a domesticated landscape impoverished by exploitative human action is undeniably a constructive act, so long as we realise that such a landscape remains largely a human product, in terms of both genesis, as suggested by Conn and Conn in Chapter 3, and execution, as argued by Mozingo in Chapter 10. While such restorations contribute significantly to our quality of life, the ongoing obliteration of what is wild, both within ourselves and within the world as a whole, is a separate issue of transcendent importance, as

Conn and Conn emphasise; and the attempt to restore the damaged wild – as a "gift back to nature," as R. France put it at the recent Harvard "Brown Fields and Gray Waters" Conference – is therefore an altogether more ambitious undertaking that raises fundamental questions about the limits of human capability. In pursuing the differences between these two types of human action, I will be questioning whether we have the capacity to repair wild nature, or whether the only gift we can make to the natural world is the gift of freedom from our cognitive maps and instrumental powers. In doing so, I hope to trace the conceptual, political, and epistemological boundaries of what is possible through restoration – a task I have organised in terms of six somewhat overlapping reservations.

Reservation I: Externalising Cognition

According to C. Lasch and other critical social theorists, a defining characteristic of the modern, narcissistic mentality is our withdrawal from any larger context – temporal, ecological, geographical, or spiritual – capable of transcending our technologically-inspired belief system.[1] Consequently, instead of living within a world that greatly exceeds and is 'other' to this belief system, the world tends to be experienced as an *embodiment* of it. Within a narcissistic culture, then, whatever qualities might exist beyond our cognitive or discursive powers appear inconceivable or even nonexistent. As a result, even when we intend to act in ways that are consistent with an 'objective' natural order that exists beyond human influence, we unwittingly reshape the world to fit our preconceptions. This should alert us to the possibility that restoration can slide into becoming a covert *extension* of industrialist processes rather than a *correction* of them, as briefly touched upon by Conn and Conn in the preceding chapter.

To the extent that we are colonised by industrialist ideology, rationality and science *seem* to provide a potentially complete explanation for events; so there is often a glossing over of the distinction between the world itself and the ways we think about the world. In other words, we come to believe that the world is 'really' made up of molecules, species, basaltic rock, and so on rather in the same way that we see society as made up of welders, vegetarians, and swing voters. Such descriptive terms relate more or less accurately to reality, but they aren't *identical to* it; and they leave out what doesn't fit with our cognitive frame. To think and act purely in such terms is therefore unwittingly to substitute the map for the territory. 'Restoring biodiversity', for example, may not amount to rejuvenating the complex web of interconnections in an original ecosystem, just as 'liberating' a country may not amount to recreating a workable social system.

Of course, as William Jordan points out,[2] experience of attempting to restore wild areas can and should expand cognitive boundaries; but while we remember our successes, there is a danger that our failures, which may have a good deal more to teach us about the discrepancies between our models of the world and the world itself, tend to be literally and figuratively 'buried'.

What we think of as restoration of wildness, then, may actually perpetuate a concealed domesticity that is inherent in our cognitive assumptions. Like J. Rodman, I am suspicious of 'revolutions' that are "led by people who perpetuate in their character the authority structure of the old regime."[3] To take one of Rodman's examples, to restore the San Bernadino National Forest by replanting it with smog-resistant Ponderosa and Jeffrey Pine may be to perpetuate a humanised nature rather than to restore a wild one. As he points out, our domestication of nature and "our division of ourselves into a 'human' part that rules or ought to rule and a 'bestial' part that is ruled or ought to be ruled are by now so hopelessly intertwined that it seems doubtful that we could significantly change the one without changing the other."[4] From this perspective, the assumption that we humans are 'in control' of what is happening to the world appears as a dangerously hubristic delusion that disguises the extent of our own colonisation by technique. As Rodman continues,

> "Descartes' depiction of beasts as machines was followed by the proliferation of mechanistic models of man; Marx's indictment of capitalist industrialism for treating human workers as machines is followed by Harrison's and Singer's indictment of factory farming for creating the monstrosity of 'animal machines'; the *Natural Resources Journal* is followed by the *Journal of Human Resources;* and Darwin's projection onto nature of a model derived from man's 'domestic productions' (plant and animal varieties created by artificial selection) now returns to haunt us as the prospect of the genetic engineering of human beings by human beings, as the literal fulfillment of the metaphor of domestication."[5]

Even when we seem to be behaving independently, we may be blind to the larger industrialist frame that subtly permeates and constrains the choices available to us. Given that notions such as 'natural', 'wild', 'organic' and 'individual' have been widely harnessed to sell the products of the commercial world, and that the experiences these notions refer to are very likely to reach us through media such as television,[6] opposition to industrialism and the intention to remedy its effects may themselves have been incorporated within industrialism. Our behaviour and experience, therefore, may well be most deeply compromised where we least expect it, even permeating our intention to remedy the effects of industrialism.[7] We are blind to these influences on our behaviour because they have become part of our own personality structure, so that we 'naturally' act in accordance with them. The danger, then, is that what we perceive as restoration of the external world may, at least in some cases, be more soberly be seen as an expression of technical mastery. The attempt to uproot an undesired present and to replace it by a slice of historical time arbitrarily selected from the 'past' should, therefore, be viewed with suspicion by those who, like myself, share Rodman's concerns. Like the apparently benign activities of white explorers donating modern artifacts to native tribes, our good intentions may carry with them structures that are ruinous to the wildness we attempt to restore.

Consciousness exists within a context of unconsciousness,[8] and so tends to be blind to the deeper swells and currents of the natural order. Domestication and the loss of the wild reflect the fragmentation and collapse of these larger, imperceptible contexts as much as any more measurable loss of 'biodiversity'; and as we intuitively recognise, simply adding the relevant species and stirring is not an adequate recipe for the restoration of wildness, although it may be a start in this direction. Since the narrow beam of consciousness is a poor tool for the reconstitution of healthy ecological structures, it is important to recognise that our restorative efforts will *necessarily* be partial and incomplete. To talk of 'restoring' wilderness incorporates the same casual hubris as our talk of 'growing' tomatoes or 'having' babies: in each case, our language focuses too much on what we do, and not enough on those taken-for-granted processes and intelligences that are embodied in the natural world itself.

To summarise, the notion of narcissism warns us that instead of recognising whatever is outside and beyond individuality as different from and separate to ourselves, we tend to assimilate it to our own psychological worlds; and our failure to distinguish between our own conceptual structures and those of the natural world leads us to reproduce the former even when we intend to recreate the latter. Such a narcissistic perspective is not simply an individual failing, but is rather characteristic of the whole social fabric in which we unconsciously participate. If we increasingly experience our own lives as detached from historical and natural cycles, is it surprising that our actions and creations should be similarly detached?[9] While such an approach may 'work' within the reductionist realm of industrial production, the natural world seems to operate in terms of a sort of multiple contextual embedding, so that one natural process always exists within others; and to omit such systemic complexities from our 'restorations' may be to omit exactly those characteristics which define the world as wild rather than artifactual. The danger, then, is that the world we 'restore' is not a wild world at all, but one of our own making; and like Pooh and Piglet, we 'find' in it those entities that we have unwittingly put there ourselves.

Reservation 2: The Denial of Natural History

Part of what may be lost through this substitution, I suspect, is the sense of a *temporal ecology*. Consciously, the present moment supersedes and replaces the past; and we apply this emphasis on the present in important ways such as the discounting of the future that is taken for granted in economic policy, and in assuming the irrelevance of the past. In general, history is viewed not so much as a *foundation for* the present, or integrated into it, but rather something that we strive to *replace by* the present, or else 'recreate' technologically, like Jurassic Park. Cognitive time, in other words, is linear and reversible, like a videotape.

While our ecological awareness teaches us that the natural world can't just be conceptualised in terms of a 3-dimensional space containing discrete entities in certain locations, our understanding of time remains simplistic and Newtonian. There is a danger, I think, that we take our temporal patterns from cognitive rather than ecological time, seeing situations as reversible in a Piagetian sense: that is, an ecological operation, like a cognitive one, can simply be reversed, so that an undesired present state can be replaced by one that existed at an earlier time, as if the undesired state had never existed.[10] In accordance with A. Light's views in the next chapter, I want to try to move away from this understanding of time as a constant, continuous dimension, existing over and above the events and processes that populate it, toward a conception of ecological time that is more intricately interwoven, incorporating the past within the present in complex ways. Such a conception of time is consistent with that of many indigenous societies: for example, R. Ridington, referring to the Dunne-za of northern British Columbia, points out that "historical events happen once and are gone forever. Mythic events return like the swans each spring ... [They] are true in a way that is essential and eternal."[11]

Time, for example, can in some ways be understood as cyclical. Thus we say that the seasons follow 'cycles'. But we would do better to say that the cycles of time are a cognitive expression of seasonal changes. Our experience of time, in other words, should be understood as *modelled on* natural processes rather than natural processes 'following' a pre-existing cyclical pattern. We mistake the abstraction, the reification, for the reality, making nature secondary to cognition, and forgetting the whole history of the emergence of cognition. And while we can recognise linearities and circularities, nature will also embody other more complex 'times' that are imperceptible to us. Mapped onto a conscious understanding that recognises only the simplest temporal schemes, many natural changes – such as spawning patterns that follow a lunar cycle[12] – appear 'random' to us.

Now, this linear and reversible understanding of time, while it 'works' in the domesticated world, can only be applied to the natural world by violence. We can take an old car engine and recycle its parts, melting them down so that they become raw material for another engine that will contain few traces of its previous history. But ecosystems, unlike industrial artifacts, are only *incompletely* understandable in these terms. True, the molecules that once constituted a squirrel may eventually become part of a Douglas Fir; but there are other ways than this in which the past may be embodied in the present, which itself embodies the future. An ecosystem contains within itself the geological strata, the growth rings embodying the effects of drought and fire, the patterns of tree growth that originate in the form of nurse logs, the eroded canyons, and the forms of intelligence that have co-evolved over millions of years: all these incorporate the past within the present, not as passive background or raw material, but as active, structural components.[13] Geological strata may have been laid down millions of years ago; but their form and other properties continue to affect the present and to interact with later events in the genesis of the present. Natural processes, then, build on and incorporate the past: they do not simply supersede or reverse it. And is it not this sense of embodied history, of gradual evolution through time, of continuous natural process, that is part of what we value about wildness? When we gaze down into the depths of the Grand Canyon, is not part of what we experience the inarticulate, somatically registered awareness of our immersion in the hugeness of history – an awareness that stills rational calculation and mutely draws us towards what lies beyond the cognitive realm?

This interwovenness of time is generally characteristic of living systems. Although, for example, we humans in the industrialized world consciously maintain the linear conception of time, we know from experience – whether or not we take psychoanalysis seriously – that the past can live in us and influence our behavior in the present. This is a criterion that differentiates the organic from the mechanical world: while we may sooner or later be able build a cyborg that in many ways perceives and thinks and behaves like a human, integrating present behavior with the past implies a very complex, *organic,* relation within temporal structure. In the film *Blade Runner,* E. Tyrell, President of the cyborg-producing Tyrell Corporation remarks that without such an integration between past and present "there is something missing". "Memories", replies Deckard; "You're talking about memories . . . ".[14] Just as events and images from birth onwards may in multiple ways determine our behaviour in some present situation, so also in ecosystems events and situations that occurred millions of years ago (say, geological events), and events that occurred in the last week (say, a beaver felling a tree), and much that happens in between will all find themselves part of present reality. Cognition tries to separate these temporal relationships, to say that we live only in the present; and that the relationship between past and present is a linear, deterministic one.

Part of what our narcissistic perspective loses sight of, according to Lasch, is "the sense of belonging to a succession of generations originating in the past and stretching into the future."[15] Consciousness emphasises individual experience in the present moment rather than in terms of our participation in an ontological fabric that stretches away from us temporally, geographically, ecologically, and culturally. And yet, to focus on just one of these dimensions, nature is about *process* as much as any frozen moment of natural history. This prioritisation of the present is built into our theories, too: psychology, for example, often understands the mind as consisting of 'strata', with a recently evolved cortex not so much integrated with, but dominating and superseding the earlier, 'primitive' parts of the brain.[16] Similarly, Freudian psychoanalysis sees conscious 'rationality' as having the power of a "dictator" over the more 'primitive' impulses of the id, and in particular, coordinating the "attack on nature" (and, by implication, on the past) that Freud and many of his followers saw as the pathway to a more fulfilling life.[17] And in parallel to the colonialist processes that it sometimes resembles, restoration of nature often follows destruction in much the same way that the ostensibly benign assimilation of natives to the dominant culture follows the overt violence of their conquest. Just as native groups may be encouraged to recreate the rituals and dress which used to be part of their culture, reinventing them as tourist attractions regardless of their location within temporal and cultural ecology, so a 'wild' landscape can be reinvented as devoid of temporal structure, in terms of its present, economic, significance. 'Wildness', in such cases, rather than standing as an alternative to, and a critique of, economic life, is redefined as a component of it.

In nature as in personal life, history is often viewed as a means of *understanding* the present, rather than as something that is embodied in it and in the 'deep structure' of what exists in the present. Note how this makes history part of *cognitive* structure, internal to ourselves, rather than *natural* structure, 'out there' in the world. Often we conflate these two meanings of 'history', so that when we use the term it is unclear whether we are referring to the course of events within which the present is rooted or to our *understanding* of these events. Rather than recognising history as something that is embodied **in** the geology and ecology of a landscape, contributing to its value and meaning, we may well see it as a sort of temporal grid of equal intervals that we place over nature in order to catalogue and order its apparent chaos – a point that is further explored in the chapters by Light, Mills, and Spelman.

The effects of conceptualising time in this way ripple outwards to affect and define our entire relation to what is 'outside' us. This makes these effects harder to identify, since they appear to us as a 'natural' part of the context that surrounds us. For example, if history is part of a cognitive realm rather than part of the natural world, then a landscape can be defined and recreated simply through its *present* material and biological reconstitution: its *historical* constitution becomes irrelevant. Consequently, as human action reshapes nature, so the historical process that once defined the natural world is eliminated, and even apparently 'wild' nature converges with its industrialist representation.

If carried out ineptly, restoration runs the risk of mapping this cognitive reality on to the natural world, perpetuating the cognitive dream of power: to reconstruct the world. We say: "This is the way it can be defined, in the present"; rather than: "This is the history it is part of". And such temporal discontinuities are paralleled by spatial ones: cognition separates 'urban areas' from 'restored areas', 'forest' from 'prairie'. But nature recognizes *transitions* rather than discontinuities: and there is an integration between whatever exists on each side of a transition, whether temporal or spatial. A zone of transition, in fact, is often a place where ecological relations are richest: think of riparian zones, for example. The sort of boundary that exists between a 'restored' area and an urban one is alien to nature, as it demarcates two zones that are in many ways unrelated to each other. While there are clear *distinctions,* for example, between predator and prey, or bee and stamen, or riverbank and river, or between the states of an ecosystem at different times, they nevertheless have rich and sophisticated *relations.* The cloth of nature is a continuous one, in other words, even if its pattern contains clearly distinguishable parts. The patterns of the industrialist world, in contrast, tend to be more fragmented: that is, they have diversity but less integration. If the differences between these two types of pattern, and temporal patterns in particular, are not adequately recognized, then our 'restoration' of wild areas will simply reproduce the compartmentalized and fragmented character of current subjectivities.

Whitehead, in the epigraph at the beginning of this paper, is suggesting an *ecology of time*. This ecology is poorly understood: how certain aspects of the past – the floods, the blowdowns, the fires – contribute creatively to the present and are incorporated within it. Consequently, we sometimes behave as if we can turn back the clock and reinstate some previous condition. This temporal insensitivity is part of the de-struction of the world; and it is as significant as the forms of destruction that we usually indicate by the term 'ecological'. Intuitively (although not rationally), we recognize that *history is part of present structure*, not a superordinate realm of cognitive 'understanding' that hovers above physical reality. If we don't acknowledge this, then we are likely to encounter the sort of difficulty pointed out by Rodman:

> "Some modifications seem difficult and paradoxical to reverse. The buffalo herd in Stanley Kramer's film *Bless the Beasts and Children* thunders out of the pen, released by the daring efforts of a group of heroic boys, only to stop and graze peacefully on a nearby hill, allowing themselves to be rounded up and imprisoned again. Elsa, the Adamsons' pet lioness, 'born free' and then tamed, must be laboriously trained *(sic!)* to become a wild predator before she can be safely released. Of such ambiguous stories is the mythology of the human condition in the 'post industrial' age composed."[18]

An authentic landscape is one that embodies its entire history: as in a human life, it will express and incorporate within the present past events and traumas, together with the processes through which they have healed. And just as repressed events express themselves symptomatically in a human life, so too in nature. Near to the part of Derbyshire where I live, Shipley Park is a 'restored' area that used to be a large coal mine. Now a mixture of woods and grassland, the soil is so compacted by the heavy machinery used in the 'restoration' that even small amounts of rain lead to extensive waterlogging and the appearance of boggy areas. While we may try to camouflage history, nature always integrates it; and such divergences between cognition and reality are likely to be expressed symptomatically.

What these ideas move towards, then, is a conception of restoration as the return of ecological integrity, – 'ecological' being understood as having to do not only with the biological relations between entities, and between entities and wider context, but also psychological, historical, and cultural relations between all these. Both restoration and human psychological healing might be seen as occurring not so much through the excising of an undesired state, together with the exotic species or experiences that go with it; but rather through their simultaneous and reciprocal reintegration within the larger structure of the natural order. In our human lives, this sort of meaning-making occurs when a problem or emotion is transcended through its integration into a larger realm of meaning such as spirituality or mythology. Applied to the natural world, such a restoration is also a self-restoration, since it requires that we recognise those parts of ourselves that exist beyond the cognitive, extending into the world 'outside' ourselves. As R. Brooke points out, we realize our spiritual nature not so much by finding it within us, but by "the world's revelation as a temple",[19] by resonating with and responding to the spiritual qualities of the world. It is, of course, a facet of narcissistic culture that all this needs to be spelled out and theorized, as opposed to simply being taken for granted as part of the essential nature of ourselves and of the world.

Reservation 3: The Denial of Natural structure

If we unwittingly equate cognitive and natural structures, then not only will we impose our categorical description of nature on to the world, but we will ignore those qualities that *aren't* recognized by this description. If we make this mistake, then it will appear that if nature is to possess any order and meaning, then these qualities will have to be *imposed on the land* rather than *recognized in it.*

As an example of the sort of thing I mean, I'd like to refer to a lively and provocative paper by B. Hull and D. Robertson, in which they argue that "ecosystems [are] transitory assemblages of biotic and abiotic elements that exist (or could exist) contingent upon accidents of environmental history, evolutionary chance, human management, and the theoretical perspective one applies to define the boundaries".[20] Nature, according to this viewpoint, is the result of essentially random processes - - at least until the infusion of order and meaning from 'human management' and a 'theoretical perspective'. The implication is clear: if nature is to possess any meaning and order, human intervention is not only permissible, but *essential!* Hull and Robertson argue that "because many possible natures exist, *which* nature is chosen to serve as the goal of restoration requires imposing human values and preferences for one time period and one set of initial, perhaps random, conditions".[21] This seems problematic for two reasons: firstly, because the emphasis on "human values and preferences" denies *nature's* preferences, and secondly, because the suggestion that the initial conditions may be "random" denies the possibility of inherent natural structure, particularly *temporal* structure – a point I will return to later.

These assumptions echo colonialist ideologies: 'cultureless' natives are supposed to *need* our education and socialization, just as 'empty' wildernesses *need* filling with 'useful' plants so that they become 'productive'. The general pattern is that we have the understanding, intelligence, and solutions; while *they* are passive, disordered, or primitive. Furthermore, according to Hull and Robertson, this is nothing new: humans have 'managed' nature for thousands of years, so that "like it or not, nature is now a human artifact."[22] Referring to Denevan's influential paper on "The Pristine Myth",[23] they suggest that North American nature has for "perhaps 10,000 years" been heavily influenced by human action; and restoration, therefore, simply continues this benign human influence. "Native agriculture and commerce were extensive", argue Hull and Robertson, "and Native Americans transformed the landscape with cities, roads, hunting, and agriculture".[24] Thus we are asked to believe that the 3.8 million individuals who, according to Denevan, populated the *whole* of North America in 1492, used fire, bows and arrows, and stone axes to produce a landscape as fully humanised as roughly 100 times as many individuals having access to the full arsenal of industrial techniques. The upshot of their argument is that the notion of a wild nature beyond human influence is simply a romantic fantasy; and restoration doesn't have to pay attention to anything that lies beyond the boundaries of what we can conceptualise, since there is nothing that lies beyond these boundaries. Conveniently, then, the boundaries of the natural world are supposed to coincide precisely with those of human cognition.

Views such as this assimilate complex natural situations to the requirements of logical argument for clear categories with no overlap. In the example above, the relevant categories are A). pristine nature untouched by human influence; and B). human artifacts. If 'wilderness', so the argument goes, can be shown to be even slightly influenced by human action, then since it cannot belong to category A, it must belong to category B. Consequently, wilderness areas are 'human artifacts'. Such arguments may be logically flawless (if we overlook the dubious premises), but the natural world seldom follows this principle of categorical distinctiveness, and embodies nuances, resonances, and relations that logic cannot cope with. If we lose sight of the distinctions between the ways we categorize the world and the world itself, then not only will we be unable to distinguish entities such as wilderness areas and urban parks, but we will also fail to see the similarities between entities that we place in different categories, such as humans and nonhuman creatures.

There is a similar glossing of differences in the statement that a "preference for non- or pre-human nature, the state of nature existing before any human contact, requires a normative judgment that humans ... are bad. But what is objectively bad about human-induced environmental change?"[25] Well, in order to unpack this question, I think we need to differentiate between industrial and non-industrial forms of human change. At the most basic, molecular, level, the industrial world does indeed use the same elements as the natural world; but at any higher level it usually embodies quite different forms of organization. The biosphere contains no natural process, for example, that produces steel, or PCB's, or polyethylene; or that depends for its power on the burning of fossil fuels. Large-scale natural processes are typically built on complex interactions between lower-level ones: for example, animal life, including our own, ultimately depends on photosynthesis, pollination, and the interaction of species at many levels. Industrial processes tend to ignore higher-level natural structures and processes such as ecosystemic organization, viewing nature simply as a source of raw materials for production processes that are mostly alien to the natural world. And although there are exceptions in both directions, it is generally true that while tribal societies have been highly integrated into their natural environment, industrial societies have tended to *replace* their natural environment with a built or agricultural landscape. Consequently, the claim that "a preference for pre- or non-human nature ... requires a normative judgment that humans are bad" glosses over differences of quite fundamental importance between industrial and tribal societies, and can be understood as an attempt to make wilderness preservation appear misanthropic. The essential distinction, I suggest, is not between 'human' and 'natural' change, but between 'natural' and 'industrial' processes, both of which may involve human action; and this latter distinction, I think, is crucial in assessing restoration projects.

Some theorists allege not only that wilderness has been progressively humanized over the centuries, but that wilderness is *in principle* a part of the humanly constructed world! The view that the world is already "a human artifact"[26] can be seen as another facet of the increasingly narcissistic social ethos in which we live, reflecting "a loss of belief in the meaningfulness or even the reality of the external world"[27], and leading to the view that "'reality' itself is a product of the activity of our imagination"[28]. This allows us a remarkable degree of flexibility in how we describe and modify the natural world. There have been claims, for example, that "wilderness . . . is as easily found in the city as in the vast rain forest";[29] and that wilderness

> "is not a primitive sanctuary where the last remnants of an untouched, endangered, but still transcendent nature can for at least a little while longer be encountered without the contaminating taint of civilisation. Instead, it

> is a product of that civilisation, and could hardly be contaminated by the very stuff of which it is made. Wilderness hides its unnaturalness behind a mask that is all the more beguiling because it seems so natural".[30]

If there was never any such thing as undomesticated nature, then clearly we don't have to worry about *losing* it; and so we have a free hand to design, restore, and alter the world in line with our cognitive models, according to human 'values and preferences'.

Here we see a convergence between certain restorationist views and those of social constructionists: while some restorationists, as we noted above, argue that the world is the product of human *physical* intervention over many years, constructionists argue that it is the result of *discursive* action.[31] In both cases, the world is seen as the product, rather than the context, of human action, and a 'wild' nature independent of such action either hasn't existed for centuries or else is impossible in principle. An ideology that begins from such a viewpoint has no power to restore wild nature, but only power to consummate the congruence between the world and our models of the world – a consummation that is achieved not by our growing into the world or by learning about it, but instead by reducing it to fit our preconceptions. Thus the narcissistic dream is fulfilled; and rather than being a source of otherness and diversity that we grow into, the world becomes a mirror reflecting our unexamined premises.

Such views suggest a closet Cartesianism that sees nature as just 'stuff', without structure or direction; and it carries the implication that any restoration has to be carried out solely by us, as the only intelligent entities. If ecosystems are "transitory assemblages of biotic and abiotic elements",[32] then as we noted above, whatever order they may contain has to be the result of human intervention. Such views of landscape embody a concealed reduction that is embodied not only in its physical degradation, by also in the restoration that attempts to reverse this degradation – an ontological reduction from an evolving, vital landscape capable of healing itself to an area of passive material without any such capabilities. Just as the genocide of indigenous populations was preceded by their conceptual reduction from intelligent, cultured, subjective beings to 'savages', in a parallel way we conceptually reduce nature to primitive biological entities existing within a passive landscape. The industrialized world, regrettably, has a history of such chronic blindness to structures that we don't recognize. To take just one example, the sophisticated songs of the Aranda, as noted by Strehlow in his formidable account of central Australian aboriginal culture, were described by one 'authority' on aboriginal society as "merely a collection of sounds and cannot be translated. There have no actual meaning, but are merely a means of expressing such music as there is in the native mind."[33] Colonialism's legitimation by such prejudices seems embarrassingly naive to us today. Could our assumption that nature consists merely of chance 'assemblages of biotic and abiotic elements' be equally embarrassing to future generations? Such prejudices are routinely applied to the natural world: as T. Ingold points out, while the foraging and hunting behavior of animals is often regarded as "worked out for them in advance, by the evolutionary force of natural selection," and ascribed to 'instinct', comparable behaviors in humans tend to be theorized in terms of consciously formulated 'cognitive strategies'.[34] This, in turn, is related to our understanding of 'intelligence' as necessarily located in an abstract realm that is separate from, and 'above', the material world. Thus, the aerodynamicist's calculations, supposedly, demonstrate 'intelligence'; but the swallow, which embodies this intelligence in its flight, does not.[35]

The stakes are high; for if such assumptions are wrong, then some types of ecological restoration can be reframed as ecological destruction, in the same way that subjecting Native American children to 'Western' education, with the wisdom of hindsight, appears as cultural destruction. We might do better to adopt the 'precautionary principle', and assume that a healthy nature embodies processes that are imperceptible to our cognitive powers, and that it has the capacity to heal itself in ways that we can't always foresee – as our experience of nature tends to confirm, in any case. Such healing may, of course, be slow and unpredictable; but then is it not paradoxical to 'restore' wild nature by imposing an industrialist time-scale and pattern on it? A truly restored landscape would be one that embodies not merely the present and the past, but also the temporal process that weaves past and present into a meaningful history. Such integration, as T. Morrison suggests, occurs in a way that resonates with human meaning-making, incorporating the past into the present in creative ways:

> "You know, they straightened out the Mississippi River in places, to make room for houses and livable acreage. Occasionally the river floods these places. 'Floods' is the word they use, but in fact it is not flooding; it is remembering. Remembering where it used to be. All water has a perfect memory and is forever trying to get back to where it was. Writers are like that: remembering where we were, what valley we ran through, what the banks were like, the light that was there and the route back to our original place. It is emotional memory – what the nerves and skin remember as well as how it appeared. And a rush of imagination is our 'flooding'."[36]

Just as, in Morrison's view, writing is a natural process that tries creatively to integrate past and present, incorporating 'emotional memory' as a valid part of the past, so an adequate restoration would restore not just the 'biotic and abiotic elements', but also the temporal ecology that a purely materialist account ignores. Remembering is not just the factual recording of past events; it is also *making sense of such events within the context of the present;* and the history of a landscape, like that of a human life, is more than a sequence of random, unrelated events. Morrison points out that the usual process of historical analysis records the 'facts' such as dates, names, and so on, but erases the entire subjective life of the people involved. Writing about the experiences of slaves, she argues that

> "If writing is thinking and discovery and selection and order and meaning, it is also awe and reverence and mystery and magic. I suppose I could dispense with the last four if I were not so deadly serious about fidelity to the milieu out of which I write and in which my ancestors actually lived. Infidelity to that milieu – the absence of interior life, the deliberate excising of it from the records that the slaves themselves told – is precisely the problem in the discourse that proceeded without us. How I gain access to that interior life is what drives me ... It's a kind of literary archaeology: on the basis of some information and a little bit of guesswork you journey to a site to see what remains were left behind and to reconstruct the world that these remains imply."[37]

The crucial distinction for Morrison, then, "is not the difference between fact and fiction, but the distinction between fact and truth. Because facts can exist without human intelligence, but truth cannot. So ... I'm looking to find and expose a truth about the interior life of people who didn't write it (which doesn't mean that they didn't have it)."[38]

Nature too, forced into slavery, didn't write its history (which doesn't mean that it doesn't have one); and similarly, reproducing a set of biological conditions doesn't add up to an authentically restored landscape. In Morrison's terms, in order 'truthfully' to reproduce a complete nature, we have to do more than factually reproduce the physical conditions of its existence – the appropriate species, the right topsoil, and so on.[39] A restoration that is merely 'factually' correct will not be 'true', since it will faithfully reflect our cognitive models of the world rather than the world itself. A more authentic form of restoration will be less concerned with imposing 'human values and preferences', and more attentive to nature's own healing processes, sympathetically embodying those attributes that are on the very edge of our capacity to recognize, such as those that Jack Turner refers to as 'presence' and 'aura'.[40] Aldo Leopold, too, recognized that wild nature embodies more than our cognitive models suggest. German forests, he lamented, lack wildness. "Forests are there ... Game is there ... [but] there is something lacking. ... I speak of a certain quality [wildness] which should be, but is not found in the ordinary landscape of producing forests and inhabited farms."[41] Such notions should induce in us a certain humility in assessing our capacities to restore landscapes.

Reservation 4: Restoring Nature, or Restoring Industrialism?

If we believe that damaged wilderness can be restored, as opposed to being allowed to heal itself, then this may have the unfortunate effect of making humanly caused damage seem less unacceptable. This is an important point, but because it is widely recognized[42], not one that I will dwell on. It is, however, one particular expression of a more general danger: that restoration may sometimes be used not so much as a way of repairing nature, but rather as a way of repairing the integrated web of beliefs, technologies, and forms of 'rational' consciousness that together allow capitalism to flourish. We forget the strip mines hidden away in remote areas of an Indian reservation; or the Indonesian sweatshops where our trainers are made; or the clearcut forest camouflaged by the 'beauty strips'. The industrial system, as T. Weiskel points out, "has a constant interest in our amnesia".[43] So-called environmental disasters such as the grounding of the Exxon Valdez can perhaps be better regarded as *capitalist* disasters, since they remind us of what we forget: that the 'rational', reasonable, humane appearance of capitalism is the cosmeticized exterior of a violent and destructive system. The real environmental disaster is the 'business as usual' of capitalism, which is usually covered up by a veil of concealment, euphemisms, and geographical distance; and there is a danger that restoration may be employed to mend a tear in this veil, rather than to rectify or prevent the destruction itself. Restoration of areas affected by oil slicks, for example, seems to have achieved limited success in saving wildlife. What it does is to convert the *visible* devastation of oil-covered rocks and dying seabirds into the *invisible* and superficially more acceptable devastation caused by cleaning with powerful detergents and high pressure hot water.[44] In attempting to reverse the effects of human action on the natural world, we may unwittingly be perpetuating the underlying assumption that problems can always be solved through consciously-formulated human action. In such cases, restoration serves the public relations needs of capitalism and quiets our own individual guilt and unease, rather than protecting the natural world.

As A. Light and E. Higgs suggest, "the context in which a restoration physically occurs (the economic, political, and social spheres around it) is crucial in determining its political role in the broader culture".[45] A seemingly benign activity, when viewed from a broader perspective, may reappear as a part of the process of destructive assimilation. This reflects a more general trend in which social criticism increasingly seems to accept the inevitability of, and to take place within, the 'global market', retreating toward an unconvincing celebration of the choices available *within* capitalism. Some critical social theorists now argue that consumer choice heralds the 'deep democratization' of society, since it foregrounds 'the authority of the consumer' as we choose between brands of fizzy drink or breakfast cereal.[46] Like the relativistic retreat from engagement with political and environmental realities that often pervades 'postmodern' academic stances, such views embody an implicit withdrawal from the world, reaching gratefully for the scraps of choice thrown from the capitalist table rather than insisting on a full menu of alternatives. If we emphasize the correction of environmental damage in selected localities while ignoring the larger context beyond these localities, we are in effect giving up on the natural world as a whole, accepting its fragmentation into a patchwork of disconnected areas within an overall capitalist frame; in which case restoration becomes a subdivision of capitalist activity. This situation is not simply an unfortunate side-effect; rather, it faithfully expresses a cognitive stance in which the world is *already* made up of 'separate' areas, demarcated by clear 'boundaries', each containing 'individual' entities, catalogued according to certain recognized 'variables'. This stance embodies an ideology from which those subjective dimensions that express the interconnectedness of the world have already been filtered.

Restoration, then, has the potential to be *either* destructive *or* supportive of wild nature, depending on political and other considerations. If, for example, an area is restored in order to attract tourists, or to lend a 'green' image to a multinational corporation, then the overall, long-term effect on wildness may be negative. The issue of whether or not restoration is beneficial cannot, therefore, be decided simply by studying in isolation the particular area concerned – an isolation that is itself part of the process of destruction. Rather, restoration is necessarily part of a political ecology the aims of which can seldom be grasped by consciousness.

Again, there is an analogy here – or, better, a continuity – with culture. A culture that has been 'restored' and prevented from changing is a dead culture, one that is the outcome of other people's cognitive expectations and images. A living culture, on the other hand, is one that adapts creatively to present contexts, resisting and adapting in ways that are never entirely predictable by consciousness. Native peoples adapt in their own way to being forced to live in the industrialized world; and this is more deeply authentic than becoming relics of the past, frozen in time, putting on performances for tourists – as has happened to a number of tribal groups across the world. Similarly, a nature that is mummified within a cognitive image, however benign this image may seem, interferes with rather than facilitates the process of natural adaptation and evolution. Such parallels between culture and nature are more than metaphors; for in a healthy world, culture and nature are overlapping and continuous structures, so that human life grows out of and extends the biological world.[47] It is the essence of the contemporary tragedy that these two worlds have been set in conflict with each other, crippling both. But the answer to this conflict is not some facile 'integration' of nature and culture; for such an integration, being based on our inadequate notions of nature, would almost certainly become an assimilation. Rather, we need to preserve enough of nature until at some future time an authentic integration becomes possible.

Reservation 5: Confusing Cognitive Limitations with Natural Limitations

Since *conservation* involves a 'hands off' approach rather than an interventionist one, its effects won't be crucially dependent on the accuracy of our cognitive models of the world. In *restoring* nature, however, any shortcomings in our cognitive models will directly impinge on the restored area. But what do we mean by 'accuracy'; and *which* version of nature do we intend to restore, given that nature is constantly changing? It is at this point, I think, that some restorationist accounts go off the rails.

Firstly, consider Hull and Robertson's statement that "today, nature is understood to be constantly changing, often in random and unpredictable ways."[48] However, while *randomness* suggests the absence of ecological order, *unpredictability* simply implies that whatever order exists is beyond our cognitive capacities to recognize. The importance of this distinction is illustrated by the well-known case of the Canadian lynx population, which varies dramatically, showing frequent crashes and recoveries with no apparent pattern. If, however, a 'phase diagram' is plotted, expressing the relation between population levels at particular time intervals, an elegant three-dimensional figure known as a 'strange attractor' appears[49], showing that behind the apparent disorder and randomness lies a remarkable and unsuspected order. While such natural systems are *unpredictable*, then, they are certainly not *random;* rather, they wander endlessly around the attractor, never reproducing any previous combination of conditions, but never escaping from the submerged logic of the attractor. To a human observer, such natural systems will appear to be changing in 'random' ways, their lawfulness hidden by the limitations of our cognitive capacity. The conclusion we should draw from this is that our inability to perceive patterns in nature may be just that – a human inability – not a confirmation of the absence of patterns. If so simple a pattern as the lynx population appears 'random' to us, what chance have we got of recognizing more complex ones involving the interaction of many species? What this suggests is that part of what we should be restoring is *temporal structure:* it may be that nature is more adequately characterized in terms of *processes* rather than *states,* and widely differing states may still be part of the same process. Consequently, returning a landscape to a particular *state,* and maintaining this state through human management, can be understood as a sort of ecological taxidermy rather than as a genuine restoration of dynamic natural process – a problem also recognized by Spelman in Chapter 7. Furthermore – and here's the rub – apparently similar states may not lie on the same attractor, and so in ecological terms may diverge dramatically from the process we are trying to restore.

States, however, are more accessible to our cognitive capacities than are processes, particularly nonlinear processes; so we find it convenient to understand the world in terms of states, confusing this understanding with the world itself and projecting it on to the world. Consider, for example, Hull and Robertson's statement that since "multiple definitions of ecosystem boundaries exist, [then] there is no way to unequivocally define the boundaries of an ecosystem and thus no value-free definition of ecosystem health. The boundaries will reflect the conceptual system used to describe them and hence will reflect the values and ideals embedded in that conceptual system."[50]

However, the fact that we find it difficult to *define* the boundaries or the health of an ecosystem doesn't mean that they don't exist; nor does it mean that the boundary or health of an ecosystem depends on our cognitive choices. The authors go on to assert that "using the health of an ecosystem as a goal for restoration and management is problematic for several reasons: 1) defining an ecosystem is problematic, 2) ecosystems are not closed systems or organisms, and 3) health does not offer clear management prescriptions."[51]

These claims, I suggest, confuse questions about the existence or non-existence of entities such as 'ecosystem health' with questions about our ability or inability to recognize and understand them. If "there is no value-free definition of ecosystem health", this doesn't mean that the term 'ecosystem health' has no correspondence to anything in nature; it simply means that our cognitive capabilities aren't up to the task of recognizing, defining, and theorizing the complexities of ecological health. Furthermore, if our conceptions of ecosystem health don't "offer clear management prescriptions", we can see this as an indication that we don't have sufficient understanding to intervene constructively, rather than as an invitation impose our own assumptions and ideals on the natural world. Particularly telling is the description of environmental decision making as "a tournament of competing conservation agendas" in which "stakeholders compete [in order] to advance their value systems". These 'stakeholders' are "ecological scientists, professional environmental managers, and involved citizens."[52] Note that nature itself is not seen as involved except as a passive recipient of our attention – a non-involvement that is further emphasized by the statement that "the trials and tribulations of restoration and management projects boil down to a debate about which nature we want and why".[53]

Such statements illustrate the serious difficulties that even such a sincerely well-intentioned activity as ecological restoration can get into if it accepts conventional assumptions about human superiority over and separateness from a passive, value-free 'nature', and confuses our own cognitive limitations with actual properties of the natural world.

Reservation 6: The Dissociation of Cognition and Feeling.

In modern industrial society, cognition, especially as applied by science, tends to be somewhat dissociated from feeling. Cognition involves a certain distance and detachment from whatever we're thinking about, whereas feeling is inherently connective; so an emphasis on cognition suggests an impoverished relationship with the object of thought. I want to explore the significance of this by referring to R. Ryan's paper "A People-Centered Approach to Designing and Managing Restoration Projects."[54]

Ryan differentiates between 'active', 'passive', and 'conceptual' attachment to natural areas, finding that people in the first of these categories, who were physically involved with the area through activities such as walking, cycling, and running, "were more in favor of minimal management, reducing management interventions and letting nature take its course".

On the other hand, "staff and volunteer restorationists expressed a more conceptual attachment; they were attached to a particular type of natural landscape such as a prairie rather than to a specific place", and tended toward "positive attitudes ... toward more intensive management strategies that promote native species (i.e. cutting exotic trees and brush, burning, using herbicides) ...

". According to Ryan, "this lack of place dependency suggests a substitutability of natural areas that is not shared by the other users."[55]

Now, this would seem to suggest that the attachment felt by these restorationists was to an *idea* (of an ideal prairie, for example), rather than being a *physically felt* attachment to the particular *place:* in other words, this attachment exists within the cognitive realm rather than the natural world. Such feelings would seem to parallel, in the social sphere, the sincere emotional commitment held by some individuals to ideals such as 'justice' or 'human rights', which may or may not be associated with a felt attachment to any particular people.

While such a stance is not *necessarily* problematic, history offers all too many examples of utopian projects and movements that were conceptually pure and humane, and yet in practice produced horrific consequences. In order to avoid such dire results, cognition needs to be moderated by other faculties involving feeling, empathy, and a spiritual sense; which is simply another way of saying that we need to behave like whole human beings. As a practical prescription, perhaps we should spend at least six months getting to know a place before we begin uprooting trees or spraying herbicides. It is, in fact, quite recently and reluctantly that we have begun to appreciate the importance of emotional factors even where relations between *humans* are concerned: our tendencies to reduce life to an assortment of biochemical mechanisms often blind us to those affectional structures which are nevertheless of overwhelming importance in our lives. Only since the Second World War have we recognized a syndrome in which young children who were adequately fed, clothed, washed, and kept warm, but who were starved of contact with other humans, often simply died.[56] Perhaps it is not surprising that tribal peoples, who invariably have a deeply emotional and spiritual connection to their land, also frequently die when removed from it.[57] Rather than ignoring these emotional dimensions, perhaps we should recognize – as does Mills in Chapter 6 – that a healthy relation to land can never be a solely conceptual one, and that it needs to extend into those rather poorly articulated areas of experience in which we are not merely detached Cartesian thinkers, manipulating a passive world, but are emotionally involved in a way that challenges assumptions about ourselves, about the world itself, and about how we should interact with it. An adequate approach to restoration requires that *our primary allegiance is to the world itself, rather than to our cognitive models of the world.* Such an approach to restoration is always, also, a *self-restoration,* since it requires that we restore those parts of ourselves that exist beyond the cognitive. In doing so, we heal a further split: that between humanity and nature, since those subjective dimensions that lie 'beyond the cognitive' are precisely those that reintegrate us within a fully felt nature, opening us to resonances that make impossible the usual scientistic stance of the 'objective' observer.[58]

Recontextualizing Restoration

If restoration sometimes seems to slide into the same assumptions about our dominance over the world as the industrialist processes it attempts to correct, the approach that I have outlined above is one that reintegrates us with the world, shifting the emphasis away from restoration of the landscape towards a restoration of ourselves to full subjectivity. There is also a certain humility implicit in our recontextualisation within the natural order – a joint recognition both of the complexity of the land and the limitations of our own cognitive apparatus.

Consider one practical example: Glen Canyon Dam. To physically remove the dam would involve enormous effort and energy expenditure, and a good deal of environmental damage. In a century or two, few traces would remain; few indications of the ethical dilemmas that it posed, the decisions reached, the passions generated. And given that fossil fuels will have been largely exhausted, perhaps our distant descendants will decide to build a new dam, unaware of the history of the previous one. We should remember Santayana's dictum that those who are ignorant of history are destined to repeat it: if we remove, deny, and cover up the consequences of our actions, rather than letting them exist as monuments to our decisions, there is a danger that our mistakes will be repeated by our descendants.

The alternative is to recognize that we are *part* of history, natural as well as human; and that we cannot *not* be incorporated into it. We cannot stand outside history, to halt or reverse the historical flow. The attempt to remove ourselves and our actions from history rests on the fantasy that we are separate from the rest of the world, that we can use our technological power both to sustain our absurdly affluent lifestyles *and* to put things back the way they were. It is a cognitive dream of having our cake and eating it, of exploiting the Garden of Eden and then restoring it; of being both part of the world and having power over it.

If left alone for a few hundred years, Lake Powell will have filled with silt; the dam will have begun to leak; and a new era of erosion will have begun. The shoreline will have become a riparian zone that is likely to be a haven for wildlife. Archaeologists of the future will examine the remains of massive concrete structures, the rusted turbines, a few dozen sunken houseboats; and the dam will become part of the history of human power, folly, and hubris. Even more ancient finds will occur: old pots preserved in the silt, petroglyphs revealed by falling water levels. Our era will have become part of the history of Glen Canyon rather than a reversible interruption to that history. It will stand as a monument to the human attempt to control the natural world, and people will ponder the follies of the Great Fossil Fuel Age. In other words, it will enter our mythology, and so participate in the evolution of our relationship to the earth, affecting both what we do and how we experience our own lives in the context of the natural world. If the notion of progress has any meaning, it is that we learn from experience – from the plagues, the Auschwitses, the Hiroshimas – which then becomes part of the foundation of our culture. And one of the main lessons of the past millennium, it seems to me, is that the unfettered allegiance to pure thought freed from feeling and spirit and intuition needs to mature into a balanced appreciation of our many faculties and the diverse relationships they can support. The creative inaction – a "letting go", in William Jordan's phrase – that results from this stance is not a paralysis, but rather a willingness to participate in rather than grasp control of the natural world. And the final implication of these musings, and one that leads us into the next section, is that the necessary restoration is not one that we impose on some part of the landscape, but rather the restoration of our own psychic integrity that comes about through the rediscovery of our relationship to and membership in the landscape around us.

PART II.

Relationships

Chapter 5

Restorative Relationships: From Artifacts to "Natural" Systems

Andrew Light

It is an old wag among environmentalists that humans have become disconnected from nature. The culprits for this conundrum are various. If it is not our addiction to technological enticements then it is our life in big cities which alienate us from our "earthen elements." The presumed result of this disconnection is that we do not respect the land anymore and turn a blind eye to the environmental consequences of our collective acts of consumption and pollution. Various bits of evidence are produced to prove this point – mostly anecdotal – such as the claim that many city-dwellers, when asked where their food comes from, will respond blankly, "from a grocery store."

What is the curative for this ailment? Surprisingly, it is not that we should send urbanites out to the factory farms, county-sized feed lots, or flavor factories in New Jersey, which actually put most of the food on the shelves of neighborhood markets. It is instead usually suggested that we should send people to wilderness areas, that we should become more connected with nature in the raw, as it were. E. O. Wilson's "biophilia" hypothesis is a good case in point. Defending a sociobiological account of why humans are innately attracted to living things, Wilson suggests that this connection is best realized in the residual attachment of humans to wild nature. This grounds a claim that the most important task at hand is to focus on "the central questions of human origins in the wild environment" (Wilson 1992).

It is probably unfair to suggest that Wilson thinks that we should all go to the wilderness in order to be better connected with nature, and implicitly, to then become better people. There are many others though, such as D. Abram and H. Rolston III, who make similar such cases and do argue that we are better people if and when we are connected to wild nature (see Light 2001). An alternative view however is that it is much more important to connect people with the natural systems in their own back yards and public places where they do live rather than striving to engage them with the environments of their prehistoric ancestors.

There are many reasons that I would make such a claim. One might be the healthy skepticism that has evolved in the past fifteen years over what is meant by "wilderness" at all by scholars such as W. Cronon and company (1996). Another would be an argument that development of human lifestyles which wind up being better for other critters and larger natural systems do not necessarily depend on encouraging an active respect for nature as a moral subject in its own right. In fact, I think we are more likely to get sustainability through changes in infrastructure than changes in environmental consciousness (see Light 2003a). But at bottom it is simply not true that visiting wilderness will necessarily make everyone care more about nature, or come to regret their "disconnection" from it and the consumption patterns engendered by that alienation any more than visiting the Louvre or the Museum of Modern Art will necessarily make one interested in the preservation of great works of art and develop a disdain for schlocky forms of pop culture. It is no doubt correct that knowledge of something – be it art or nature – can encourage appreciation and even value of it, but exposure to something does not necessarily get us knowledge of it, and though they are no doubt connected, development of taste does not necessarily make for a coherent or consistent moral psychology.

But rather than further developing those arguments here I will assume their plausibility and investigate another topic. What if there is something to this worry about our disconnection from nature absent the more extreme suggestions that are offered to cure us of this problem? What if it is true that we would be more respectful of natural systems, and more interested in maintenance of their integrity or health, if we came to care more about them because we did think of them as part of our lives? My sense is that such questions need not necessarily lead us down the road to a family trip to Yellowstone. The nature that most of us should encounter is much closer to home, in parks, for example, or just in gardening in our backyards.

My central claim in this chapter will be that one way in which we can find ourselves in a closer relationship with nature is through the practice of restoration of natural ecosystems, quickly becoming one of the most influential forms of contemporary environmental management and landscape design. As I have argued at length elsewhere, one of the more interesting things about ecological restorations are that they are amenable to public participation. If we give a chance to members of a local community to help to restore a stream in a local park then we offer them an opportunity to become intimately connected to the nature around them. There may be more important bits of nature for people to be connected to as they are ones that they can engage with often, even everyday, rather than only thinking of nature as residing in far flung exotic places set aside for special trips. It is like coming to appreciate a good set of family photos, some of loved ones long past and some still with us, and not worrying too much that our homes are not filled with original works of art or that we get to visit those places where such art is on display.

But if restorations offer us the opportunity to become reconnected with nature (though they may not necessarily solve that problem with the grocery store answer) what kind of relationship do we have with the things that we restore? This is an important issue to take up given the question of what will motivate the production, maintenance, and preservation of restored landscapes. Our family photos, no matter how much we love and cherish them, are ultimately not as valuable as a Da Vinci or Pollock. Or even if they are as valuable to us, if we are the last of our family line then no one may care about the safekeeping of those treasured heirlooms once we pass. The "Mona Lisa," in contrast will likely always have someone to care for it and protect it absent some fairly extreme circumstances. It already stands in relation to a community of those who appreciate it and hence it will be taken care of. Are restored landscapes ultimately more like the products of the great masters or my father's 35 millimeter camera? To try to get some answer to this kind of question I will first summarize the importance of restoration work today and the deflationary philosophical response to it. I will then offer a series of arguments which try to describe the kinds of relationships that are possible through restoration work and how they are both alike and different from the kinds of relationships which we can have with other things produced by humans.

My hunch however is that a local restored environment and a wilderness park are actually more closely connected than many of my fellow environmentalists would think. As I will argue below, if we come to care about the places closer to home then we will probably think more about the consequences of our lifestyles on their kin, the more exotic and dramatic landscapes. This may not in the end reconnect us to the primordial places where humanity evolved, but it might do something much more important: help us to find a way to live as better environmental citizens.

I. Restoration, Participation, and Sustainability

Ecological restorations can range from small scale urban park reclamations, such as the ongoing restorations in urban parks across the country, to huge wetland mitigations. In all cases, restorationists seek to recreate landscapes or ecosystems which previously existed at a particular site but which now have been lost (e.g., wetlands, tall grass prairies, and various riparian systems). On two indicators of the importance of environmental activities – number of voluntary person hours logged on such projects and amount of dollars spent– restoration ecology is one of the most pressing and important environmental priorities on the national environmental agenda. For example, the cluster of restorations known collectively as the "Chicago Wilderness" project in the forest preserves surrounding Chicago, would attract at their height some 2500-3000 volunteers annually to help restore 17,000 acres of native Oak Savannah which have slowly become lost in the area (Stevens 1995). The final plan for the project is to restore upwards of 100,000 acres. As for financial commitments, the restoration of the Florida everglades begun during the Clinton administration will come in at over $8 billion making it one of the largest single pieces of environmental legislation in history.

As a scientific practice restoration ecology is allied with such academic disciplines as field botany, conservation biology, landscape ecology and adaptive ecosystem management. But as an environmental practice most restoration in the field is undertaken by landscape architecture and landscape design firms (see Chapter 10 by Mozingo and Chapter 11 by Ryan). Restoration sites must be carefully planned and designed as they are actively created rather than only identified and protected as existing natural areas. A casual reader of the leading journal for practitioners in the field, *Ecological Restoration,* will quickly see that its back pages are dominated by advertisements for landscape architecture firms specializing in restoration work and by universities seeking to attract students to programs of study in landscape architecture. Some past presidents of the Society for Ecological Restoration, such as T. Clewell, head prominent landscape architecture firms.

Recognizing that successful restorations must bring together various allied fields in environmental science with the design strengths of landscape architecture demonstrates the inherent interdisciplinarity of this activity. But in addition to the scientific and design questions at the heart of restoration work, which have received substantial attention in the literature, there are also ethical issues which bring to light competing priorities for any given project. Unfortunately these ethical issues, and the dilemmas they sometimes present for restorationists, have been woefully under explored by environmental ethicists.

When restoration is taken up by environmental ethicists, the results are mostly negative. While there are some notable exceptions (see for example Gunn 1991, Rolston 1994, Scherer 1995, and Throop 1997), the most influential work by environmental philosophers on this topic, surely that of E. Katz and R. Elliot, have largely consisted in arguments that ecological restoration does not result in a restoration of "nature," and that further, it may even harm nature considered as a subject worthy of moral consideration (Katz 1997, Elliot 1982 and 1997).

These criticisms stem directly from the principal concerns of environmental ethicists, namely to describe the non-human centered (nonanthropocentric) and non-instrumental value of nature (see Brennan 1998 and Light 2002a). If nature has some kind of intrinsic or inherent value – or value in its own right regardless of its use to anything else – then a wide range of duties, obligations, and rights may be required in our treatment of it. This is much the same way that we think about the reasons we have moral obligations to other humans in many ethical systems. Kant's duty based ethics argued that each human has a value in and of themselves such that we should treat them never only as a means to furthering our own ends but also as an end in themselves. But one immediate worry is that if nature has a value in comparable terms then a discernable line must be drawn between those things possessing this sort of value and those things which do not have this value and hence do not warrant the same degree of moral respect. Such a demarcation line is critically important, for if it cannot be established then the extension of moral respect beyond the human community might result in an absurd state of affairs where we hold moral obligations to everything around us. If that were the case then perhaps I am doing something unsavory at the moment by merely using the pen I am writing this chapter with only to fulfill my own ends. Thus, the demarcation line designating natural value in a moral sense must distinguish between "nature" and non-natural "artifacts" or realms of identifiable "nature" and "culture."

One problem with restored landscapes for both Elliot and Katz is that they can never duplicate the value of the original nature which has been lost and which restorationists seek to replace. The reason restorations cannot duplicate the original value of nature is that they are closer on the metaphysical spectrum to being artifacts rather than nature, especially when the latter is understood as an object of moral consideration. Restorations are the products of humans on this account; they are merely artifacts with a fleshy green hue. For Elliot, their value is more akin to a piece of faked art than an original masterpiece.

But such a view is the best case scenario for restorations on such accounts. Katz argues that when we choose to restore we dominate nature by forcing it to conform to our preferences for what we would want it to be, even if what we want is the result of benign intuitions of what is best for humans and nonhumans. Katz has argued that "the practice of ecological restoration can only represent a misguided faith in the hegemony and infallibility of the human power to control the natural world" (Katz 1996).

In part however Katz has softened his position in this regard, responding to recent criticisms that he thinks remediation is often our best policy option: ". . . the remediation of damaged ecosystems is a better policy than letting blighted landscape remain as is" (Katz 2002). His reasoning here is that blighted landscapes are no longer really natural and hence our interaction with them cannot necessarily count as an instance of domination of nature. Such a view should sanction most restoration since very little of it, if any, is aimed at interfering with pristine landscapes. (This of course begs the question of whether restoration can ever lead to domination since we generally don't try to restore landscapes that haven't been damaged. No matter though. I will leave this worry for the moment.) But immediately after offering what may be his strongest positive claim yet about restoration, Katz repeats one of his now familiar criticisms: ". . . once we begin to adopt a general policy of remediation and restoration, we may come to feel omnipotent in the manipulation and management of nature. And thus we will create for ourselves a totally artifactual world" (Katz 2002). Harking back to his earliest criticisms of restoration, Katz still insists that the practice of restoration will encourage us to develop more "pristine" areas under the assumption that we will now think that we can always make up for the harm we have done to nature through restoration.

Unfortunately, such claims have received much attention by restoration practitioners. As a result, many of them have come to the unfortunate conclusion that philosophy is largely unhelpful in sorting out future directions for restoration practice. So reliant is such work on difficult to defend and often

ot only a restoration of nature, but also of the human cul-
nature (this idea is developed in Light 2002b).
s particular moral advantage of restoration requires that
in these projects be actively encouraged. Ecological
oduced in a variety of ways. While the Chicago restora-
high degree of public participation, others have not.
in these various projects has been a result of their differ-
ity. Dechannelizing a river in the Everglades will be a task
Army Corps of Engineers and not a local community
torations that could conceivably involve community par-
gh do not (even when there is no obvious liability issue at
h already involve community participation do not utilize
nuch as they could. Each restoration therefore represents
y to link a local public with its local environment and
onstituency devoted to the protection of that environment
vardship rather than law.
iestion though is what kind of relationship is produced by
tored landscape? If we start from the nonanthropocentric
nd Elliot then it is difficult to see restorations as anything
cts. If we do not start from that perspective then whether
oes produce "nature" or not is immaterial to the question
ve direct moral obligations to it. We can't. There is how-
that produces a range of alternatives for us (another mid-
d by Jordan and Turner in Chapter 1). In the remainder of
to give a more specific defense of the moral basis for pub-
uggesting different ways in which we can conceive of our
with restored natural systems. This argument will build
on this topic (especially Light 2000b) but also go beyond
exercise is to show that the restoration of the human rela-
is possible even if ecological restorations are culturally pro-
efully, even if one is skeptical of the artifact worry, unpack-
h a relationship in more detail will help to give us addition-
ze public participation in restoration whenever possible.

with Objects

of Katz and Elliot to ecological restorations. Common
im that restored landscapes are not part of nature, they are
t's assume that they are correct in holding this view if only

tedious arguments about the metap
empathize with this response.

Because of this situation I have
come the bad rap of philosophers w
philosophical criticisms of Elliot a
2000b) and then moving forward
issues involved in this practice. As
potential for restorations to serve
more actively involved in the enviro
tial for work on restoration project
and stewardship (see especially Lig
further argument than I have spac
been that a direct, participatory rel
and the nature they inhabit or ar
encouraging people to protect na
rather than trade off these enviro
development. If we have a strong
probably less likely to allow it to
ships however does not require th
of agent in and of itself that can
means that we must come to care
because it has a place in our lives v
for one expression of this idea).
the land around us is to actively
offers us the opportunity to do ju

Importantly however, the value
cation. In the case of restoratio
demonstrated to get us better rest
tionships with nature suggested a
Chicago restorations suggests that
are more likely to adopt a benig
toward nature as a result of such
The reasons are fairly obvious: p
about the hazardous consequence
they learn in practice how hard it
aged. There is thus a strong emp
can serve as a kind of schoolhous
participatory restorations create o

restorations bec
tural relationsh

But captur
public particip
restorations car
tions have invo
Partly the diffe
ing scale and co
for an outfit lik
group. But ma
ticipation often
stake), and some
that participatio
a unique oppor
arguably to creat
bound by ties of

A still pressi
interaction with
perspective of Ka
other than mere a
or not a restorati
of whether we ca
ever a middle gro
dle ground is exp
this chapter I will
lic participation b
possible relationsh
on my previous w
it. The goal of th
tionship with natu
duced artifacts. H
ing the quality of s
al reasons to maxi

2. Relationship

Recall the objectio
to both views is a c
artifacts. For now,

restorations become not only a restoration of nature, but also of the human cultural relationship with nature (this idea is developed in Light 2002b).

But capturing this particular moral advantage of restoration requires that public participation in these projects be actively encouraged. Ecological restorations can be produced in a variety of ways. While the Chicago restorations have involved a high degree of public participation, others have not. Partly the differences in these various projects has been a result of their differing scale and complexity. Dechannelizing a river in the Everglades will be a task for an outfit like the Army Corps of Engineers and not a local community group. But many restorations that could conceivably involve community participation often enough do not (even when there is no obvious liability issue at stake), and some which already involve community participation do not utilize that participation as much as they could. Each restoration therefore represents a unique opportunity to link a local public with its local environment and arguably to create a constituency devoted to the protection of that environment bound by ties of stewardship rather than law.

A still pressing question though is what kind of relationship is produced by interaction with a restored landscape? If we start from the nonanthropocentric perspective of Katz and Elliot then it is difficult to see restorations as anything other than mere artifacts. If we do not start from that perspective then whether or not a restoration does produce "nature" or not is immaterial to the question of whether we can have direct moral obligations to it. We can't. There is however a middle ground that produces a range of alternatives for us (another middle ground is explored by Jordan and Turner in Chapter 1). In the remainder of this chapter I will try to give a more specific defense of the moral basis for public participation by suggesting different ways in which we can conceive of our possible relationships with restored natural systems. This argument will build on my previous work on this topic (especially Light 2000b) but also go beyond it. The goal of this exercise is to show that the restoration of the human relationship with nature is possible even if ecological restorations are culturally produced artifacts. Hopefully, even if one is skeptical of the artifact worry, unpacking the quality of such a relationship in more detail will help to give us additional reasons to maximize public participation in restoration whenever possible.

2. Relationships with Objects

Recall the objections of Katz and Elliot to ecological restorations. Common to both views is a claim that restored landscapes are not part of nature, they are artifacts. For now, let's assume that they are correct in holding this view if only tedious arguments about the metaphysical status of nature that it is easy to empathize with this response.

Because of this situation I have been trying over the past few years to overcome the bad rap of philosophers working on restoration by first answering the philosophical criticisms of Elliot and Katz on restoration (Light 2003b and 2000b) and then moving forward to explore a different aspect of the ethical issues involved in this practice. As suggested above, my focus has been on the potential for restorations to serve as opportunities for the public to become more actively involved in the environment around them and hence in the potential for work on restoration projects to encourage environmental responsibility and stewardship (see especially Light 2000a and 2002b). While it would take further argument than I have space for here, the foundation of my claim has been that a direct, participatory relationship between local human communities and the nature they inhabit or are adjacent to is a necessary condition for encouraging people to protect natural systems and landscapes around them rather than trade off these environments for short-term monetary gains from development. If we have a strong relationship with the land around us we are probably less likely to allow it to be harmed further. Forming such relationships however does not require that we come to see nature itself as some kind of agent in and of itself that can be dominated like another human. It simply means that we must come to care about the land around us for some reason because it has a place in our lives worth caring about (see next chapter by Mills for one expression of this idea). One way that we might come to care about the land around us is to actively work it in some way. Ecological restoration offers us the opportunity to do just that.

Importantly however, the value of public participation needs further justification. In the case of restoration, participatory practices can be empirically demonstrated to get us better restorations because they create the sorts of relationships with nature suggested above. Sociological evidence focusing on the Chicago restorations suggests that voluntary participants in restoration projects are more likely to adopt a benign attitude of stewardship and responsibility toward nature as a result of such interactions in restorations (see Miles 2000). The reasons are fairly obvious: participants in restoration projects learn more about the hazardous consequences of anthropogenic impacts on nature because they learn in practice how hard it is to restore something after it has been damaged. There is thus a strong empirical basis for the moral claim that restoration can serve as a kind of schoolhouse for environmental responsibility. At its core, participatory restorations create opportunities for public participation in nature;

to avoid the dicey question of whether such a distinction between realms of nature and culture is at all coherent. Implicit in this assumption is that our relationships with artifacts are not as strong as the relationships we could have with natural systems once we have come to recognize that natural systems have a direct moral value that should be respected. But what may be overlooked on such views, which may provide some helpful middle ground, is that artifacts can bear meaning in a normative sense in a way that does not degenerate into some kind of occult view. At the very least, objects can be the unique bearers of meaning for relationships between humans that holds strong normative content, and in that sense we can interact with them in ways that can be described as better or worse in a moral sense (see also Chapter 7 by Spelman).

There are lots of examples of how we can relate to each other in better or worse ways through objects. Some may find trite the examples that come to mind – the political meaning of flags for instance (I was terrified as a young Cub Scout to let the American flag touch the ground simply because I was told that it was wrong). But it would seem hard to deny that objects can stand for the importance of relationships between humans such as is the case with wedding bands. There may even be some argument to be made that we should respect some objects in their own right. To be more precise, I would maintain that we can be lacking in a kind of virtue when we do not respect objects in some cases, especially when such objects stand for the importance of relationships we have with others, or, as in the case of justifications for historical preservation, respect the creations of those who have come before us. Note that I use the term "virtue" here rather than stronger language of obligation because I don't think we have obligations to objects themselves in the same way we have moral obligations, for example, to people. A word of caution though before proceeding To further unpack these intuitions we should probably set aside arguments for historical preservation, for example, of landmark buildings, since they may represent a special case of moral obligation. Such claims are less about personal relationship we have through things than instances of collective obligation to people in the past or to respect the aesthetic integrity of some things. Instead I want to focus for now on more everyday examples.

One way to explain the value of everyday things is to consider the case of the destruction of an object that stands for a relationship in some way. The unthinking destruction of an object that bears the meaning of some relationship between individual humans reflects badly on the person who destroys that object. Consider the problem of replacability and replication of objects that are special to us. I have a pair of antique glasses of which I am very fond because they were the glasses that my maternal grandfather, Carmine Pellegrino, wore for much of his adult life. The glasses are a combination of a set of lenses which were no doubt reproduced at the time in large quantities and stems that he fabricated himself. The stems are nothing fancy, just bits of steel wire that he bent and shaped – he was a coal miner not a jeweler – but it is important to me that he made them. If you were to come to my apartment and drop Carmine's glasses down the incinerator shoot and then replace them with a pair of antique glasses from a shop near by, then I will justifiably claim that something has been lost that cannot be replaced. Further, paraphrasing one of Robert Elliot's famous examples about ecological restoration, if you make an exact replica of the glasses and fool me by passing them off as the original, then, while I may not feel the loss, I will nonetheless have suffered a loss of some sort unbenounced to me. And if I find out that you tricked me with the replicas then I will justifiably feel regret and then anger!

Another way of interpreting the meaning of my grandfather's glasses in my life (though one that I am not necessarily wedded to) would be along something like aesthetic grounds: the normative weight of destroying the glasses could be understood in narrative terms. Carmine's glasses play a role in my life such that they are part of the narrative of my life, or the story that I and others might tell about me. I came to have these glasses through part of the story of my life and they bear meaning because of the part that they play in that story. Counterfactually, they would bear different meaning for me if they were part of someone else's story of their life. So, to replace or replicate the glasses "interrupts" the narrative role that they play in my life. One might conclude then that their meaning is then limited to narrative meaning in this rarefied sense. But if this is true, is the harm that someone would do to me by destroying the glasses fully captured in terms of their interruption of the narrative of my life? This suggestion seems dubious. Mere aesthetic content for an object does not necessarily imply normative content of the ethical sort and the potential harm done to me by destroying the glasses seems normatively significant in a moral sense as well.

Surely the moral harm that may be done to me in this case is parasitic on the value of having been in a relationship with another person and not simply in some quality which inheres in the object itself. Still, the object does play an irreducible role in this story – it is a unique entity that evinces my relationship with my grandfather that cannot be replaced even though the relationship in this case is not only represented in this object. Still, both the relationship and the object have some kind of intrinsic value. But surely not all relationships have this kind of value and so neither do all objects connected to all kinds of relationships. How then can we discern the value of different kinds of relationships?

One possible source is S. Scheffler's work on the value of relationships. Scheffler is concerned with the question of how people justifiably ground special duties and obligations in interpersonal relationships without this only being a function of relations of consent or promise keeping. Scheffler's account argues for a non-reductionist interpretation of the value of relationships which finds value in the fact that we often cite our relationship to people themselves – rather than any explicit interaction with them – as a source of special responsibilities. So, for Scheffler:

> "If I have a special, valued relationship with someone, and if the value I attach to the relationship is not purely instrumental in character – if, in other words, I do not value it solely as a means to some independently specified end – then I regard the person with whom I have the relationship as capable of making additional claims on me, beyond those people in general can make. For to attach non-instrumental value to my relationship with a particular person just is, in part, to see that person as a source of special claims in virtue of the relationship between us." (Scheffler 1997)

On this view, relationships between persons can have value in some cases not because of any particular obligations that they incur, but because of the frame of action that they provide for interactions between persons. As Scheffler puts it, relationships can be "presumptively decisive reasons for action." While such reasons can be overridden, they are sufficient conditions upon which I or you may act in many cases.

What I find most attractive about Scheffler's argument is that it conforms to our everyday moral intuitions about relationships – for example, it does not reduce them to explicitly voluntary events – and it makes sense of why we find some relationships morally compelling in a noninstrumental way. So, relationships that we are in that we find valuable in and of themselves in this way, I will call "normative relationships," as they are the sort of thing that we can be better or worse in relation to and can provide better and worse reasons for action.

One of the interesting things about the relationships that we value intrinsically though is that most of them are symbolized in objects – wedding rings, mementos, gifts, etc. For this reason then, at a minimum, we can do harm, or more accurately, exhibit a kind of vice, in our treatment of objects connected to those particular kinds of relationships. Take for example the watch I am wearing right now as I write this chapter. This watch was given to me several years ago by my former partner's parents in Jerusalem as a way of welcoming me into their family. The occasion is cherished by me even though I am no longer in an ongoing relationship of the same kind with her or her parents. The watch is however a meaningful symbol of that event and that set of relationships. If someone were to try to take this watch from me and smash it I would have a presumptively decisive reason for stopping them that was not limited to its value as mere property but also would include its value as a thing standing for a particular normative relationship. So too if I were simply to smash this watch myself with a hammer for no reason. I would be doing something wrong in some sense relative to the intrinsic value of that set of relationships as well. To tease out my intuitions on why it would be wrong to smash the watch I need not appeal to any obligation to the thing itself but only claim that I have presumptively decisive reasons to respect the watch because to do otherwise does harm to a connection of value involving my relationship with others in which the watch plays some role. Again, it may help to think of this in terms of vice. I exhibit a kind of vice when I smash the watch. This is a minor vice, but it is a vice nonetheless. My character is lacking if I do not seem to minimally care about this object when it is appropriate for me to do so.

Does this example mean that my character is necessarily flawed if I smash the watch? No. Under some circumstances it might even be appropriate to destroy an object from a past relationship out of some justified anger over the course of the relationship. But where no such reasons exist, and the object stands for a relationship still cherished, such an action would be questionable. Someone hearing me brag about smashing this watch for no reason might justifiably hesitate in forming a relationship with me. Does this example imply that the meaning or significance of the relationship which the watch represents is lost if I smash it? Certainly not, as any object is not the primary bearer of the meaning of any relationship. Does this mean that all objects bear meaning in this way? Again, no. Just as the value of some relationships with others can be reduced merely to instrumental terms, so too the value of some objects can be reduced merely to their use or exchange value.

Now imagine that I show you a second watch that I own – a plain cheap plastic digital watch. This is the watch that I use when I go running in the afternoons so I can see how long it takes me and I can find out if my time improves as I continue to run. I actually don't remember where I got this watch. If I smash this watch then very little is implied about my character as it does not bear any meaning that has normative content that can reflect on my relations to others.

Finally on this point, if there is something to these intuitions then the meaning of objects in this normative sense can fade over time. But importantly this is not a unique property of objects since the meaning of our relationships with other persons can also fade over time. Still, recognizing that the normative content of objects can fade deserves some attention. If I find an object in an antique store, say a watch made in 1850 with an inscription from a wife to a husband in it, would it be worse of me to smash it than it was to smash the plastic runner's watch? If I can presume that this watch stood for someone else's normative relationship, even though that person was not me nor anyone that I knew, is there something better or worse about my character depending on how I treat that object? I probably do not want to think about the meaning of my treatment of the antique store watch in the same way that I would the treatment of an object which has meaning in a relationship I am in now, but I think there is something there that should give us pause. Maybe whatever the meaning of the 1850 watch is, we can imagine it as providing something akin to the reasons we might have to pay attention to the value of old buildings or the treatment of historical artifacts and why we ought to hesitate to smash them up too. Still, it also might be that we have independent reasons to try to respect such objects as well, similar to the arguments I have offered so far in this section. I will return to this point at the end.

But where does this discussion get us with respect to our topic at hand, ecological restorations? At least it gives us reasons we can build on to find value in restorations even if, as Elliot and Katz have it, they are only artifacts. On this account however their value as artifacts will also depend on how they help to mediate the sort of human relationships that are presumptive reasons for action.

3. Restoration as a Source of Normative Ecological Relationships

There are no doubt many ways to describe the value of nature. We are natural beings ourselves and so nature has value as an extension of the value that we recognize in ourselves. The resources we extract from nature are valuable at least insofar as we value the places for ourselves that we construct out of those resources as well as their role in sustaining our lives. And certainly there is something to the intuition that other natural entities and whole systems are valuable in some kind of non-instrumental sense even if we can be skeptical that this sort of value offers sufficient resources to justify moral obligations for their protection. Is there anything else?

Consider again Scheffler's argument about the value of relationships. When applied to considerations of the environment this approach resonates somewhat with the focus in environmental ethics on finding non-instrumental grounds for the value of nature. But rather than locating these grounds in the natural objects themselves, an extension of Scheffler's views would find this value in relationships we have with the natural environment, either (1) in terms of how places special to us have a particular kind of value for us, or (2) in the ways that particular places can stand for normative relationships between persons. On reason (1), certainly Scheffler would have trouble justifying the value of such relationships between humans and non-humans, let alone humans and ecosystems using his criteria, but I think there is no *a priori* hurdle in doing this especially if we can separate Scheffler's claim about the non-instrumental value of such relationships from the possible obligations which follow from them. Focusing just on the value of these relationships we can imagine having such substantive normative relationships with other animals whereby the value we attach to such relationships is not purely instrumental. We do this all the time with our relationships with pets. And why not further with nature, more broadly conceived, or more specifically with a particular piece of land? Because the value of such relationships is not purely instrumental, reciprocity is not a condition of the normative status of such relationships but rather only a sense that one has non-instrumental reasons for holding a particular place as important for one's self.

For some like Katz, the moral force behind such a suggestion would best be found in a claim that nature is a moral subject in the same or a very similar way that we think of humans as moral subjects. So, just as we can conceive of being in a relationship with other humans as being morally important, we can conceive of being in a relationship with any other non-human subject as important in the same way. Again though, this claim rests on a form of nonanthropocentrism that Scheffler, and probably most other people, would find highly contentious. And it would miss an important part of what I'm trying to argue for here: it is not only the potential subjectivity of nature that makes it the possible participant in a substantive normative relationship, it is the sense that nature, or particular parts of nature, can be "presumptively decisive reasons for action," because being attentive to such a relationship can be assessed as good or bad. If I have a special attachment to a place, say, the neighborhood community garden which my family has helped to tend for three generations, then whether I regularly visit it to put in an afternoon's work can be assessed as good or bad because of the history that I have with that place. My relationship with that place, as created by that history, creates presumptively decisive reasons for action for me in relation to that place.

The same would be true if I were in a substantive normative relationship with another person. There would be something lost or amiss if I didn't contact them for a year out of sheer indifference (for an example see Light 2000b). In such a case, my indifference could be interpreted as reason to doubt that the relationship was important to me at all. So too, something would be lost if I didn't visit the community garden for a year out of indifference. But what would be lost need not rely on attributing subjectivity to the garden. My relationship with the garden is a kind of place holder for a range of values, none of which is reducible as the sole reason for the importance of this relationship. To distinguish this kind of relationship from others, I want to call it a "normative ecological relationship," both to identify it as a relationship involving nature under some description in some way and just in case some wish to set aside for later consideration the issue of how this sort of relationship might be substantively different from other normative relationships. Critically though, because this argument does not depend on attributing something like intrinsic value to nature itself, let alone subjectivity, the metaphysical status of the object in such a relationship is not important to the justification for forming a relationship with or through it.

We should also note here that if I am in a normative ecological relationship with something this does not mean that my reasons for action derived from that relationship could never be overridden, either in the face of competing claims to moral obligations I might have to other persons or other places, or because of some other circumstances which caused me to separate myself from that place. It only means that my normative relationship to the place can stand as a good reason for me to invest in the welfare of that place. Also important is that the moral status of my relationship to such a place does not exist in an ethical or historical vacuum. If my relationship to a place has been generated out of my experience of having acted wrongly toward others at some site (let us say it is an inhumane prisoner of war camp where I worked contentedly as a prison guard) then my character can be justly maligned for so narrowly understanding the meaning of a place that has been a source of ills for others. Outside of such extreme cases though my relationship to places can exhibit the qualities that we would use to describe our relationships with others, such as fidelity and commitment.

Can ecological restorations be a source of such normative ecological relationships? It seems entirely plausible if not unassailable that they can.

Sociological research, like that mentioned above by Miles, is quite convincing on this point. In her study of 306 volunteers in the Chicago Wilderness projects the highest sources of satisfaction reported were in terms of "Meaningful Action," and "Fascination with Nature" (Miles 2000). "Meaningful Action" was gauged, for example, in the sense in which restorationists felt that they were "making life better for coming generations," or "feeling that they were doing the right thing." "Fascination with Nature," was correlated with reports by volunteers that restoration helped them to "learn how nature works" (Miles 2000). Participation in restorations can give volunteers a strong sense of connection with the natural processes around them and a larger appreciation of environmental problems in other parts of the world. Said one volunteer, "The more you know, the more you realize there is to learn," not just in terms of understanding the peculiarities of a particular restoration site, but also generating a greater appreciation for the fragility of nature in other places in the face of anthropogenic distress.

People can form important relationships with the restorations that they participate in helping to produce. No doubt, some will still demur that the things produced in a restoration are nothing but artifacts, but in this sense at least it doesn't matter. Assuming that a particular restoration can be justified for other ecological reasons – that it produces an important ecosystem service such as stemming the loss of native biodiversity in an area or even simply cleaning up a site so that it is a better habitat for persons and other creatures – the issue of whether a restoration is really "nature" is practically moot on this account. Just as in the case of the special watches from the last section, the objects produced by a restoration can be valuable in and of themselves as special things to us and as holders of important sources of meaning in our lives. This claim does not prohibit us from criticizing those restorations which are intentionally produced either to justify harm to nature or to try to fool people that they are the real thing. But such restorations, which I have termed "malevolent restorations" (Light 2000b), can be discounted for the same reasons that we would discount the attachment that people have to persons or places that are morally tainted in other ways.

As I said before, the case for the normative status of objects like my grandfather's glasses or my special watch is most clearly seen when an object in question stands for a relationship between persons. Recall that earlier in this section I suggested that an extension of Scheffler's views on the value of relationships could be similarly extended to natural objects not only because of the value such objects have to us but secondly because of the ways that particular places can become meaningful between persons (admittedly, this is a difficult distinction to draw in many cases though I do not take the flexibility of the distinction to be a problem as such). To my mind, the real power of those restorations that maximize public participation is that they create not only relationships with places but also relationships with persons as mediated through places. Following a very close third in the Miles survey measuring satisfaction among restorationists in Chicago was a category that the surveyors called "participation," understood as the sense in which participation in restoration activities helped people to feel that they were "part of a community," or that restoration activities helped them to see themselves as "accomplishing something in a group." As suggested in section one, the moral status of restorations is arguably not only as a source of normative ecological relationships but also as the place holders or repositories of normative relationships with other persons. Accordingly, as we saw with the examples in the last section, to fail to respect the integrity of those restored places which have such a status for persons, regardless of whether they really are "nature" or not, is to exhibit a kind of vice.

While data like this from Chicago is limited, anecdotal evidence from the field confirms it. To paraphrase R. Putnam (2003), public participation in restorations helps to produce a kind of natural social capital in a community. It can become one link between people that helps to make them a community and as such the products of restoration can be respected as part of the glue that holds a community together (also see Jordan and Turner in Chapter 1). Why does participation in restorations help to make stronger communities? It could be because it produces a sense of place for people helping them to lay claim to a particular space as definitive of their home (see Chapter 6 by Mills). But it could also be that for some volunteers there is something akin to the creation of a direct normative relationship with nature that is played out in something like phenomenological terms. They come to see the restorations they work in as part of who they are. Still, for others, restorations may be a source of transgenerational value: if different generations of a family work the same restoration site then it may become a material link between them akin to the material link I feel to my grandfather through his glasses. What is important though is that none of these reasons needs to be considered decisive on this understanding of the value of restorations. Like the value we may find in artifacts, the reasons that we decide to be more careful in our treatment of a thing will most likely be multiple and overlapping, mirroring the multiple reasons we have for finding the relationships in our lives important. Because the framework here is not one entailing a form of nonanthropocentrism, which would of necessity need to find a value directly in a restored landscape and give reasons why it had value and other things did not, there is less reason to come up with a single grounding for this kind of value.

Some will object that lots of kinds of participation in public projects can create this kind of value. Certainly this is true, though there is no reason why the grounds for these kinds of relationships has to be unique or why they must be embodied in one kind of artifact (be it a green one or not) rather than another. If part of the value of participation in restorations is that they create opportunities for us to be in moral relationships with each other through something that either is a part of nature, or is at least connected to other things which are natural, then other opportunities to create those kinds of relationships will be valuable as well.

The point which must not be lost though is that the potential of restorations to produce these kinds of moral relationships with places and between persons is most likely only possible when people actually get to participate in either the production or maintenance of such sites, and hopefully in both. If we see the practice of landscape architecture as a moral practice, responsible to producing positive natural values in the same way that we may see a responsibility for architecture in general to produce things with positive social values (see for example Harries 1998 and Steiner 2002), then the best restorations designed by landscape architects will involve a component of public participation in them as well (this subject is explored in Chapter 11 by Ryan). Good restorations which include this participatory component will maximize natural values by producing a set of relationships of care around such sites which will help to insure their protection and preservation into the future. At the same time, participatory restorations have the potential for producing landscapes inclusive of strong social values between persons as well. While no architectural project of any kind can do everything, maximizing public participation in restoration is at least one goal which is feasible as a mark to aim for whenever possible. As with any noble aim we give up much when we do not try to reach it at all.

Conclusions

Two points by way of conclusion to anticipate possible objections to this argument so far. If we can have normative relationships through objects then what about the case of a relationship with other humans that is made manifest through producing things which rely on the destruction of nature? If it is the case that we ought to respect artifacts because of the role they can play in our lives with each other, as material embodiments of the moral import of relationships with each other, then the same will be true for larger artifacts and entire built spaces. Many if not most of these built spaces will involve the destruction of the environment. Couldn't the social advantages of restoration then be captured just as effectively by acts of destruction of nature like public participation in a project that would drain a wetland in order to put up a shopping mall?

While certainly such a case could be made, at least two considerations would mitigate such a counterexample. First, as mentioned before, the moral status of one's relationship to a place does not exist in a vacuum. While certainly it is true that humans need to destroy parts of nature in order to live, a project involving the wanton or irresponsible destruction of nature, even if it was inclusive of community participation, would be objectionable for other reasons. Again, we need not resort to a claim that nature has nonanthropocentric value to come to such a conclusion, but can also argue that preserving rather than destroying nature is justified for more traditional, human reasons for finding nature valuable.

Second, if the value of objects as I have described it is partly parasitic on normative relationships between persons then we can justifiably ask whether the production of an object that we might stand in some relations to might also harm our relation with other persons. While I have no space here to further unpack such a claim, it is well known that some built spaces can be criticized for their role in harming the social cohesion of communities rather than strengthening it. Even if a space is communally produced, its value in the end for strengthening a community may be undercut by the product itself. To take an extreme case, if an ethnically diverse community were to pull together and voluntarily build a club that would be open only to whites, then the values of community cohesion produced through the production of the place would in the end be in contradiction with the probable effect of such an institution on the community. To be sure, some volunteer restorations may also produce disagreement and dissent in a community, especially over the question of what to do with public land. I have previously discussed such cases (Light 2000a). For now however I only want to say again that the value of doing anything needs to be weighed against all foreseeable benefits and problems that it will produce. To be sure, if we are going to take seriously the social capital produced by a restoration project then we must also take seriously claims that it diminishes social capital as well.

(2) If public participation in restoration produces normative ecological relationships, do such relationships persist over time? What is the status of a restored site once the community which produced it passes out of existence?

Such a case is similar to the example of the 1850 watch described at the end of section two. I said there that our reasons for preserving the 1850 watch might be similar to those we have for historical preservation in general which are beyond the scope of this chapter. But I also hinted that there might be independent reasons to show some modicum of respect for such objects based at least in part on the analysis provided here. A more through answer would need to be worked out, but at bottom part of respecting persons, all other things being equal, most likely also involves respecting the reasons that they have for holding the things important to them as important to them. Such a basis for respect is certainly not unassailable, but where possible we ought to respect, for example, the relationships that people hold dear as part of who they are as a person. This may entail respecting the objects connected to those relationships as I have suggested above. In the case of people who are no longer around but where the residue, as it were, of their relationships persists in the objects that outlast them, the respect owed to them through those objects is certainly diminished (both Kidner in Chapter 4 and Spelman in Chapter 7 discuss the temporal relationships embedded in restoration).

But importantly, when we stand in some kind of relationship to the persons embodied in these objects which stand for such normative relationships – where, for example, we are related to the persons who are gone – then we will have better reasons to respect those objects and even protect and preserve them where possible. Such is most likely the case of the family heirlooms and photos mentioned at the start of this chapter. We appropriately feel more compelled to take care of our own old family photo albums than we would of those we might find in a thrift store.

In the case of ecological restorations, such an approach should yield some hope for finding reasons for respecting restorations produced by communities that is even more robust than the case of family heirlooms. Because of the long-term nature of restorations, multiple generations of people, sometimes even from the same families, can participate in them. This trans-generational quality of such projects makes them more likely to be cared for in the future. But because it is unlikely that any one family can complete a restoration on its own, the communities which will stand in relation to them through the production and maintenance of such sites not only stretches into the future but will also be open and permeable to a broader community than an object important in only one family. A restored site will not be the property of one family to be passed down from generation to generation but to a larger community whose membership will change over time. Certainly, there are no guarantees. It is still possible that a restoration could be so successful that it becomes self-sufficient, requires no on-going maintenance, and that as a result the community responsible for its initial production passes out of existence. But if that is the case then hopefully other reasons for protection of the site will become self-evident, inclusive of reasons for maintaining and preserving other naturally evolved ecosystems. Whether such a site has an origin as an intentionally produced landscape will probably matter little given the other natural values that it will help to maintain.

For all of the reasons offered so far in this chapter, the moral potential of restoration ecology, even if the objects produced by this practice are artifacts, is that they can produce "restorative relationships" between persons and nature, as well as simply between persons. What can be restored in a restoration is our connection to places and to each other. As I said at the beginning, much has been made of the claim that humans have become disconnected from nature. I am not so sure how connected we ever really were. But if it is correct that we were more connected at some time then perhaps the relationships possible through ecological restorations can go far in more concretely helping to shore up those connections. I for one am very happy that we need not go too far away from home to learn this lesson.

Chapter 6

Restoring and Renewing Whole Ways of Life

Stephanie Mills

About twenty years ago I moved from the city to the country. I wanted to live closer to nature and to live more simply, to be a little more congruent[1] with my decentralist, deep-ecologist ideology. My desire had elements of romanticism, naiveté and elitism about it, to say nothing of impracticality.

Yet my aim was good, I think. I was interested in reinhabiting a life-place. As articulated by bioregionalist Peter Berg and conservation biologist Raymond Dasmann, reinhabitation "means learning to live-in-place in an area that has been disrupted and injured through past exploitation. It involves becoming native to place through becoming aware of the particular ecological relationships that operate within and around it. It means understanding activities and evolving social behavior that will enrich the life of that place, restore its life-supporting systems, and establish an ecologically and socially sustainable pattern of existence within it. Simply stated, it involves becoming fully alive in and with a place. It involves applying for membership in a biotic community and ceasing to be its exploiter."[2]

Reinhabitation need not necessarily be rural, especially in a world of megacities, where half of humanity is urbanized. But after a decade and half in a very fine city, I wanted to belong to a biotic community with more trees and wildlife than people per acre. Upon my arrival in Leelanau County, Michigan, I got my wish.

Within a few years, I became the owner of thirty-five acres of recovering rural land. According to the notes from an 1880 survey, the section of the county where I live was wooded with sugar maple, beech, elm, ash, basswood, and hemlock. The elms left no heirs, the ash are elsewhere, and of the other trees' descendants, on my property there are only a couple of little patches of second or third growth hardwoods remaining, stranded at the south boundary.

Almost as soon as it was settled in the mid-19th century most of the county, along with the rest of the Northwoods, was cut for timber. Ours fueled the Great Lakes steamer traffic. When the trees were gone, the land was farmed with limited success. The soils are light, extremely sandy. Gravel, stratified by the melting of the glaciers that once blanketed this land, is among the county's leading exports. Probably the last crop grown on this land before it stopped being a farm and became the commodity — and luxury — known as "acreage" was corn. Now four-fifths of my parcel is forested with rows of increasingly rangy Scotch pines. Planted either for erosion control or as Christmas trees, they anchor what remains of the humus that the long-lost forest produced.[3]

I am grateful to be ensconced in theses pines and the cherries, beeches, and maples arising among them. The landscape is not primal, but compared to what surrounds it, it's a howling wilderness. Yet I wonder how much longer even this secondhand version of the Northwoods can persist, given both the systemic threats to wild life such as climate change, alien species invasions, transgenic organisms, and a myriad of pervasive contaminants; and the relentless subdivision of our once-rural county.

However inexact a term "nature" may be, it generated diversity and dynamic stability far better than civilization has. It's a pity we are obliterating the example. In a wild process that moves through the generations, bacteria, fungi, plants and animals shape themselves, to place, inhabit the terrain, weave taut webs of give and take, conditioning and enriching their ground. No wild place is lacking in consequence. Small and fleeting vernal pools and puddles in the woods, where salamanders congregate to meet and mate have existential meaning for those amphibians. Places the size of biomes can teem with life, overwhelming in kinds and numbers. Two hundred years ago, the biome where I live was a vast unbroken hardwood forest whose spawn of seeds, nuts and berries sustained the fabled billions of passenger pigeons.

With such wild abundance but a memory, albeit recent, it's tempting to conclude that, give or take a few previous extinction events, nature puttered along pretty well for a few billion years before, Homo sapiens, bipedalism, the opposable thumb, and an anatomy conducing to language evolved. Throughout the millennia between hominids' first employment of the fire stick for landscape modification and our tapping into the one time only bonanza of fossil energy, human practices had taken a slow and steady toll on natural diversity and abundance. Our species is an ecological dominant, a habitat-reforming omnivore. We have the general effect of simplifying our ecosystems. The advent of agriculture, prerequisite for civilization, and then the industrial era marked points of inflection where the human population, its economic activity and ecological impacts began to grow exponentially.

Civilization, dependent on agriculture and trade, can only exist by exploiting and reducing wild terrain. This human practice of reducing natural diversity for short-term advantage has in the modem era with the accelerations provided by explosive population growth, rapid and pervasive global transportation and powerful technology brought the biosphere to the brink of, if not collapse, then drastic simplification.

Extinction is forever. What's gone is gone, and lamentation can't call it back. Yet acknowledging that industrial civilization has wrought havoc with the planet's ecosystems, their indigenous peoples, and for its ostensible beneficiaries, is a crucial step. Aldo Leopold's declaration, in his *Round River* that "One of the penalties of an ecological education is that one lives alone in a world of wounds"[4] wouldn't be quoted so frequently if it lacked resonance. Of course we must hasten to treat the wounds and help them heal, to restore such health and wholeness as we can. We must also, and of this more later, make it our larger purpose to disarm the perpetrators, forswear the wounding and found regenerative cultures.

Leopold's ecological education let him infer what was missing from the many American landscapes he'd roamed through. Thus his elegiac statement "I am glad I shall never be young without wild country to be young in."[5] He must have intuited that for its fullest maturation and flourishing, the ontogeny of the human psyche needs to recapitulate its phylogeny. By far the greatest portion of our existence as a species was spent engaging with forest edges, shores and savannas. Land health gave us our nature. Our surroundings continue to affect our mentality. As we experience, so do we think. The reality of the wild is primary, not a trope. If we cannot somehow arrest and reverse our drift into a technologically determined milieu we human beings will be living in a hard, slick mirror realm, with only artifacts in depauperate landscapes for referents.

A merely human, discursive world-as-concept is an abysmal and degrading, if truly postmodern prospect.

Suburban sprawl, road building, deforestation, farming, grazing, and the various systemic threats mentioned before are all moving the planetary biosphere and its human economies towards some simplified and exceedingly precarious mean. As the natural world and the cultural diversity it fosters are reduced to a dull sameness, a wealth of human integrity, local knowledge, skillful means and decent subsistence also is devalued and destroyed. Everyday living degenerates to getting and spending, hustling, passivity, or paranoia. Degraded places stunt growth and sap souls.

Insanely quixotic though this may be, the pre-settlement climax ecosystem remains my ideal of place, a vision to reinhabit towards. To make a prayer for the return of something more like a pre-settlement ecosystem, I have interested myself in ecological restoration. On my 35 logged-off, farmed-out replanted acres, I make the occasional move toward easing out the nonnative Scotch pines that dominate my land and encouraging the return of its original community of deciduous trees and woodland plants.

Of course even ecological restoration, a benevolent intervention in the life of the land, risks reinforcing the solipsistic view of nature engendered by the domesticated environments that civilization produces. The wild and native become contingent on civilization's none too dependable charity. Alas, all the earnest efforts to study, quantify, reconstitute, and manage the wild will be moot without an equally earnest effort at containing ourselves, our settlements and our enterprise. The root problem is not calculating the minimum number of grizzlies in a viable population and plotting the cores and corridors necessary to manage their survival, but figuring out how our species can be restored to some just proportion in the living world, how our myriad communities can establish ecologically and socially sustainable patterns of human existence.

I don't mean to argue that restoration and conservation biology and a well informed ambition to "rewild" the landscape aren't good and necessary things to do. These actions are vital to reinhabitation. However our approaches shouldn't reproduce the civilizational and industrial paradigms of centralized control and mass, rather than artisanal, production.

The penalties of ecological ignorance, which is industrial civilization's norm, are proving to be severe. Yet the natural history which is a remedy to such ignorance tends to be sequestered. To learn that the peninsula where I live had been lushly forested, indeed, belonged to the great Northwoods I had to peruse early land surveys and soil maps of the area. The primal ecology of this region wasn't a clear feature in local consciousness. Here, the woods had been primarily deciduous. Today we have a patchwork landscape of fields, suburbs, third or fourth-growth deciduous woodlands, and monocultural conifer stands like my *Pinus sylvestris.* From these remains, one couldn't infer a forest consisting of mixed stands of gigantic hardwoods, including elms and chestnuts, along with the maples and beeches and their associates, interspersed here and there with pure stands of mammoth White or Red pines.

Most folks have no concept of native vegetation and if they do, they may not understand how vital it is to preserve and restore it. The last few centuries' wide-eyed enthusiasm for moving plants from coast to coast over natural barriers like mountain ranges, and from continent to continent across oceans has utterly changed the physiognomy of ecosystems. The deliberate rearrangement of the world's flora was motivated by the quest for novel ornamental plants, faster growing trees for forestry, more vigorous forage for range lands, and new kinds of food crops. If a weed is "a plant out of place," then most of the world's landscaping, horticulture, siliviculture and agriculture consists of the introduction and propagation of weeds. Add to this the inadvertent introduction of alien species that results from trade and transport and the historic landscape is so effaced as to become a matter for research, speculation and imagination.

Yet through the attempt to learn what the primal ecosystem of a given place likely was, a reinhabitory culture can propose answers to questions like: How can we reinstate a fitting biotic community here? By following the money that was made from the commodities that primal ecosystem yielded, the question of how to build an ecological economy in a particular place can be pointed, the point being to prosper the land and lives of locals rather than, or before those of absentee owners.

Restoration, then, is a battle with entropy, working with suites of living organisms and lively human intelligence. The ultimate goal should be instating graceful contemporary forms of order and self-organization without causing disorder somewhere else. By contrast civilization, imperialist and centrist, depends on ever more extensive and intensive exploitation of its hinterland. Civilizations undermine the agrarian and subsistence lifeways of their subjects and contaminate or desertify their territories. Thus they are wont to collapse.[6] It's a likelihood to factor into the plan.

South of the section where I live, right across the state highway is a vast regional landfill. One February Sunday I made a first ski foray south beyond my acreage, across the state forest land it borders, and on out to the edge of the road. The landfill's heavy equipment, respecting the Sabbath, was silent, but its olfactory presence was pronounced. This particular Sunday the garbage moraine was emitting—wasting—methane. The miasma reached through the clones of popple and arcades of young Red pine where I'd skied, overpowering my idyllic forest fantasy. The whiff is likely to be a feature of this area for years to come. The landfill is a going concern, nowhere near the end of its useful life. Most North Americans and those who cater to them don't place much of a premium on durability of goods, and have yet to demand reusable containers and reclamation of discarded materials. Remediation and restoration of the square mile section of the county the landfill occupies, let alone a destination resort to cap it, will happen, if they ever do, in some distant future.

I do a fairly good job of minimizing my immediate output of garbage. Nevertheless I'm complicit in a system that rapes and wastes in places out of my sight. The half-dozen or so times a year I drive to the landfill to dispose of my bags of trash, I throw stuff away—but not very far away. Even though I know it's like bailing with a perforated bucket, an "easy piety,"[7] I can't conceive of not recycling. Still, to recycle may be to approach the problem from the wrong end.

Really, it's production that's the problem as much as consumption and its ejecta. Recycling in absence of a commitment to reduce consumption, and to clean up and constrain production is not unlike restoring or reclaiming degraded lands without equal and passionate action to preserve wild and free landscapes; not unlike transmogrifying brown fields and gray waters with the kind of energy intensive apparatus that designed them into existence.

Whatever memorials we make out of industrial wastelands should illuminate the full ecological as well as the social consequences of industrial enterprise. There are all kinds of lessons to be learned from the audacity and immensities of industrial landscapes. This history should not be neglected, nor

should it be romanticized. The categorical difference between the sublimity of wild nature and that of the dark satanic mills that castellated the landscape of the industrial age must be made clear. I worry that an aesthetic analogy between the wild sublime and the industrial sublime might take civilization even further down the path of imagining that the natural and the artificial, or technical are comparable.

What constitutes good in landscape design and architecture may be a question of taste and feasibility; judging the good may be done by exclusively human, political and subjective criteria. What is good for the land, conducive to land health, is perhaps more objective—if less feasible. For land to sustain the diversity of life over the long haul, it must be healthy—with soil that stays mostly in place, water cycling through earth, plants, and sky; diverse vegetative cover, pollinators abounding, predators and prey keeping each others' excesses in check and countless other symbiotic relationships thriving.

As Aldo Leopold famously put it in *A Sand County Almanac*, "A thing is right when it tends to preserve the integrity, stability, and beauty of the biotic community. It is wrong when it tends otherwise."(8) Thus the question whether the proposed reclamation or remediation project will serve artfully to enshrine, employ and sanction the wounding practices that cause brown fields and gray waters—and limitless opportunities for restoration—is primary and ethical. Of course invoking a land ethic has radical implications, could bring most of our enterprises to a screeching halt.

Be that as it may, the metatstatic enterprise of the global economy, spawned in an ethical vacuum, is bringing the Cenozoic era to a conclusion and immiserating most of humanity. A halt is overdue. It's past time, but perhaps not too late, to test our good ideas and grand designs against the land ethic and the purpose of reinhabitation In a reinhabitory culture with an integral commitment to land health, the precautionary principle ("When an activity raises threats of harm to human health or the environment, precautionary measures should be taken even if some cause and effect relationships are not fully established scientifically.")[9], if not animism or simple courtesy, would govern our dealings with natural systems. Expedience, convenience, risk management, or satisfaction of minimum requirements are paltry bases for any dignified relationship, let alone the human relationship with earth's ecosystems.

Resurfacing and regulating industrial civilization in hopes that it will function like a clean machine isn't a fitting response to the Earth's sixth great extinction crisis, which the aforementioned civilization is finalizing. Healing what remains of the biotic community rather than palliation of the symptoms of its wasting disease will require the kind of fundamental cultural and political change Berg and Dasmann call reinhabitation. While it may not be incumbent on design professionals to tackle the problem of industrial civilization head-on, it is important that all of us who hope to make a creative and life-enhancing intervention in degraded landscapes engage in some deep questioning of the causes of the degradation and of the system-wide effects of the remedies being proposed.

The system includes the human psyche, widely captivated by the idea of progress, generally intrigued by technology, and lured into hubris by technology's immense, pervasive, but partial power. Technological fixes—like the Green Revolution, nuclear energy "too cheap to meter," fiercer antibiotics, a hydrogen economy, genetic engineering, and nanotechnology have been offered as the next steps, the latest solutions to the problems that inevitably followed on the attempts throughout history to transcend our organic essence and to treat the earth as our pantry rather than our matrix.[10]

A dialectic between technology and ecology is an interesting notion, possibly worthwhile if the fundamental difference between the two is acknowledged. Human purpose directs technology. The planet's ecology, self willed, gives rise to patterns within patterns, and if there is a purpose it is impossible to know. We can only guess that it might be one greater than civilization's hurrahs. Wild nature generates all manner of fanciful phenomena like butterflies, bower birds, platypi, angel fish, bombardier beetles and slime molds. In light of the contrast between the harmless extravagance of ecology and the toxic excesses of technology we might hope that in a dialectic between technology and ecology, technology would be a good listener.

Thanks mainly to the overreaching that technology implements, civilizations crash leaving deserts in their wake. In our singular moment, because industrial civilization has gone global there's nowhere left to tum—except away from the mechanical to the organic, thereby to restore what we can of culture and place.

Restoration is an emerging, provisional discipline blessed with an artistic dimension. Restorationists devise and essay their techniques then must hope for a sufficiency of time to nurture their sites and observe their results. Ecosystems move more slowly than people, taking millennia to mature and centuries to recover from disturbance. Even more than remediation or reclamation, restoration is by nature a slow process. In all our dealings with the land, we'd do well to become more reflective in our actions and get in stride with its slower cycles.

From 1936 to 1941, the late Theodore Sperry oversaw the pioneering restoration of the Curtis Prairie at the University of Wisconsin's Arboretum at Madison. In 1991, when we met, he was perhaps restoration's most venerable practitioner. When I asked him how long it would take to restore the prairie, Sperry replied "Roughly...a thousand years."[11] It remains to be seen whether even the most exacting restoration ecology will succeed in its aspiration to restore successional processes and cradle the continuing evolution of Earth's current ensembles of life forms. With anthropogenic changes of geophysical magnitude at play, the uncertainties multiply and the odds grow long.

There are days when I wonder whether being an aspiring reinhabitor and dabbler in ecological restoration doesn't represent a hope so fantastic that it borders on denial. In absence of some fundamental, devolutionary changes in the global political economy and tempering of technological hubris, ecological restoration may prove to be a futile exercise. Even so, in their respectful, attentive, fine-grained and gently experimental approach to the once and future biotic community, the practices of restoration and reinhabitation seem likelier to yield a more durable relationship to the life of the land than the Promethean spirit of the industrial and silicon eras and the vision of an industrial, or virtual sublime. The likelihood that ecological restoration and reinhabitation will prevail seems dubious, but that's no reason not to do the work.

For the beauty immanent in the practice of restoration and the development of the craft is inherently rewarding. The verdant hope is that the work will restore land to health, engendering natural communities that will include ecologically wise human beings among their members. Reinhabitation is the cure for our ravening estrangement from the land organism. Being organisms ourselves, could we reestablish our creaturely reciprocity with our diverse habitats? Summoning our wit and will and capacity for invention, will we craft the appropriate technologies, local economies and indigenous cultures that will allow us to dwell in our ecosystems sustainably?

Restoration has different orders of meaning. Exacting work to restore natural areas to the greatest possible degree of historical authenticity is ecological restoration in its purest and most definite sense. Yet we might also be thinking about restoring and renewing whole ways of life, becoming native to our places-reinhabiting them.

Too many ethical and aesthetic travesties have been committed in the name of efficiency, progress or growth. Only by a brutal negation of the senses and of common sense and common decency could the destruction of the environment have proceeded so far. Nevertheless, our kind, having become human in them, must still possess an ability to recognize wholeness, balance, and beauty. To reclaim that inborn human sensitivity is restoration work, too.

Chapter 7

Embracing and Resisting the Restortive Impulse

Elizabeth V. Spelman

Environmental restoration constitutes but a sliver of the repair work engaged in by human beings. This much is already suggested in the previous chapters, which explored ways in which projects undertaken to repair or restore natureal sites may at the same time involve rebuilding relationships between humans and the land (Light) or rehabilitating connections among humans engaged in those projects (Mills). More broadly still, repair is in fact ubiquitous, something that we engage in every day and in almost every dimension of our lives. Indeed, we humans are repairing animals: *Homo sapiens* is also *Homo reparans*. In highlighting this fact, the aim of this chapter is not so much to focus on environmental restorative projects themselves, but rather to provide reminders of the enormous repertoire of activities attributable to *H. reparans.*[1] Such a background will enrich our understanding of the range of possibilities before us as we consider what to do, or to refrain from doing, in the face of ecological damage.

On the Ubiquity and Variety of Repair: A Brief Reminder

Perhaps the most obvious kinds of repair are those having to do with the inanimate objects with which we surround ourselves — the clothes calling out for mending, the automobiles for fixing, the buildings for renovating, the works of art for restoring. But our bodies and souls also are by their very nature subject to fracture and fissure, for which we seek homely household recipes for healing and consolation, or perhaps the expert ministrations of surgeons, therapists, and other menders and fixers of all manner of human woes. Relationships between individuals and among nations are notoriously subject to fraying and being rent asunder. From apologies and other informal attempts at patching things up, to law courts, conflict mediation, and truth and reconciliation commissions, we try to reweave what we revealingly call the social fabric. No wonder, then, that *H. reparans* is always and everywhere on call: we, the physical world we live in, and the objects and relationships we create, are by their very nature things that can break, decay, suffer damage, unravel, fall to pieces.[2]

The English language is generously stocked with words for the many preoccupations and occupations of *H. reparans:* repair, restore, rehabilitate, renovate, reconcile, redeem, heal, fix, and mend – and that's the short list. Such linguistic variety is not gratuitous. These are distinctions that make a difference. Do you want the car repaired, so that you can continue to commute to work? Or do you want it restored, so that you can display it in its original glory? Is a patch on that jacket adequate, or do you insist on invisible mending, on having it look as if there never were a rip to begin with? Should that work of art be restored, or simply conserved? Why do some ecologists favor preserving an environment over trying to restore it to an earlier state? Does forgiveness necessarily restore a ruptured relationship, or simply allow a resumption of it? What does an apology achieve that monetary reparations cannot – and *vice versa*? What was thought to be at stake for citizens of the new South Africa in the contrast between restorative justice and retributive justice – between the healing promised by a Truth and Reconciliation Commission, and the punishment exacted through an adversarial court system?

As crucial as such distinctions are, the family of repair activities share the aim of maintaining some kind of continuity with the past in the face of breaks or ruptures to that continuity. They involve returning in some manner or other to an earlier state — to the bowl before it was broken, to the friendship before it began to buckle under the weight of suspicion, to the nation before it was torn apart by hostility and war. Even though taking superglue to the bowl repairs it without fully restoring it to its pre-shattered condition, both repairer and restorer want to pick up a thread with the past. Their work appears to involve something distinctly different from the original creation of the bowl, but also from its accidental or deliberate destruction, its abandonment, or the serendipitous retrieval of its shards for flowerpot filler. In a similar fashion, there is a difference between putting a friendship back together and simply letting it hobble on, decisively ending it, or making a new friend altogether.

Repair wouldn't be necessary if things never broke, never were shattered, never frayed, never splintered nor fell to pieces – or if we didn't care that they did. A world in which repair was not necessary would either be filled with unchanging unbreakable eternal objects (a version, perhaps, of Plato's world of Forms), or a junk heap, things, people and relationships abandoned when they no longer functioned in the requisite manner. To repair is to acknowledge and respond to the fracturability of the world in which we live in a very particular way – not by simply throwing our hands up in despair at the damage, or otherwise accepting without question that there is no possibility of or point in trying to put the pieces back together, but by employing skills of mind, hand, and heart to recapture an earlier moment in the history of an object or a relationship in order to allow it to keep existing.

H. reparans has been known to take great satisfaction in exercising the capacity to repair, and pride in what the result of such exercise can do for broken-down cars, torn retinas, and frayed partnerships. Indeed sometimes we take greater pleasure in having a well-repaired object than an unbroken one, prefer (if we're lucky enough to have options) living in a neighborhood of renovated houses to one in which the buildings are all spanking new, enjoy a friendship that has known apology and forgiveness more than one protected from the risks of being rent. We seem at least sometimes to welcome the sentiment that things are stronger in the broken places.

At the same time, a voracious appetite for fixing can lead to poor judgment about what is and is not desirable or even possible to repair. Pride in our repairing abilities may push us into believing that whatever has been broken can be and ought to be fixed. Indeed, we live in an historical moment – though perhaps this has always been true – in which there is on the one hand worry that our capacities for repair are atrophying as we become habituated to throwing out broken things and moving from one imperfect relationship to the next, and on the other hand concern that we are so enamored of our reparative talents that we intervene where we should not. So we are told that "Americans, infantilized by labor-saving devices and a service industry that has put even the smallest mending or cleaning task into the hands of professionals, no longer feel at home in their own homes";[3] but also that we live in "a culture sustained by the faith that there are technical fixes for all human ills."[4] Hey, it doesn't matter if we bulldoze that wetland: we can always fix it right up later on.

Checking the Restorative Impulse

Worries about whether our reparative skills are on the one hand atrophying or on the other are exercised excessively provide a useful reminder that among the necessary skills of *Homo reparans* is the self-reflexive one of judging when, where, and whether to deploy them. Repairing or restoring or rehabilitating is not always deemed the best plan of action. We sometimes wonder whether it isn't the better part of wisdom to leave the flaws, the fragments, the ruins, alone: restorers of "Gone With the Wind" had to decide whether a flaw in the original film should "be fixed or retained as an intrinsic part of the original masterpiece";[5] a political columnist counsels her readers that "You don't have to be abused or betrayed to have a bad marriage – a marriage that cannot be fixed, even with the help of all the therapists on the Upper West Side, or all the preachers in Louisiana."[6]

A particularly provocative plea to check our reparative skills appears in architect L. Woods's *Radical Reconstruction.*[7] Woods urges us look carefully at the design possibilities to be found in buildings that have been ravaged by war, neglect, or powerful natural forces. The kind of possibilities he has in mind emerge from within the damage itself; they will be lost if a building is either restored to an earlier state or razed to make room for something new. These two otherwise quite different ways of treating damaged buildings involve what Woods takes to be a hasty and questionable retreat from the state of brokenness – a state that reveals possibilities precluded or pre-empted by a building in good repair, exhibiting the order imposed by its design and functioning the way it is supposed to.

Woods may seem to be an unlikely or unwelcome partner in the large project animating the present anthology. For in arguing that buildings in damaged condition present unique opportunities, he seems also to be suggesting that we should be less disturbed than we are by living in what Aldo Leopold famously described as a "world of wounds."[8] As we shall see, however, Woods's work is driven not by the belief that the damage surrounding us is to be cherished, but rather by worries about what is obscured by restoration as a response to that damage. And it turns out that the exploitation of brokenness he urges upon us has striking parallels in a variety of familiar human activities, each of which grows from rather than tries to undo or dispose of wreckage. The question remains, of course, whether the creation of new possibilities out of the debris is in any given case more desirable than restoration, rehabilitation, or abandonment.

Unlike "Ruinists" – those who are drawn to oases of disrepair such as Angkor Wat, the Acropolis, Machu Picchu, or for that matter contemporary industrial wastelands[9] — L. Woods's interest in damaged sites is not a matter of relishing irreparability, of taking delight or finding instruction in the reminder that even the most solid of human constructions finally fall prey to the forces of nature. Like Ruinists, however, he appreciates the ways in which historical events are palpably embedded in ruined sites. Indeed one of his worries about the razing of what may appear to some simply as ravaged, unusable eyesores is that it often involves the attempt to erase painful or unwanted memories. Such efforts, he urges, while understandable, ought to be carefully examined: far too often the architecture created in the name of starting anew offers a suffocating, "totalizing system of space and of thinking".

The early twentieth-century modernist architects, who were aligned with the ideologies of industrialization, faced the task of rebuilding an intellectually bankrupt and war-devastated culture following the war that presumably ended all wars. These avant-gardists embarked on a war of their own, employing the violence of what later would be called "urban renewal" against the presumed chaos of old cities (actually, against their nonconformance to patterns of industrialization), proposing to erase the most conceptually corrupt parts in order to build more humane cities. We should be wary, Woods urges, of those want to get rid of remnants of the old order, who want to create a *tabula rasa* on which they shall carve their design and thereby "lay claim to the future".

But Woods is equally suspicious of those who promise to reclaim the past through restoration. Restorative projects that aim to return buildings and spaces to their undamaged condition (or to the state of repair prior to their current damaged condition) erase history just as much as do the bulldozers that haul off the rubble and leave cleared, historically unmarked space: according to Woods, they represent the denial of postwar or post-catastrophe conditions; they ultimately serve "only the interests of the decrepit hierarchies, struggling to legitimize themselves finally through sentimentality and nostalgia, a demagoguery that is all too comforting and appealing to people struggling to recover from the tragedy of profound personal and cultural losses".

What, then, does Woods propose we do with sites in states of serious disrepair or ruin, if we're supposed neither to complete the road to their destruction and haul off their remains, nor try to restore them to an earlier, purportedly more desirable state of repair, and yet not simply treat them as another example of Ruins? What other options are there?

He wants us to look into the damage and be attentive to what it can tell us or suggest to us. In the wreckage of bombed-out buildings in Sarajevo, the dilapidated boulevards of Havana, the earthquake-shattered edifices of San Francisco, Woods insists, are "the beginnings of new ways of thinking, living, and shaping space". Crumbling walls, beams that have collapsed into each other, ragged slashes in floors: all are a kind of "mutant tissue, the precursor of unpredictable regenerations". Architecture constructed around such tissue is "radical," according to Woods: in producing spaces "in which you do not already know how to behave", it refuses to accommodate the functions and designs of authorities who have an interest in being able to patrol and predict behavior.[10]

It is not easy to read Woods's architectural drawings, and his substantive claims about the history and function of architecture and architects have invited the criticism that he overstates the effect of architecture on its inhabitants, and that his architecture obscenely aestheticizes war and other catastrophes.[11] But the point of trying to describe his work here, in an anthology premised on the fact that of widespread environmental damage, is his suggestion that in wounds are the seeds of possibilities precluded or occluded or in any event not easy to predict from the structure of buildings in good repair. The exploration and exploitation of such abundance is foreclosed in the rush to establish order, whether that order be dictated by the state of the object prior to the damage, as in restoration; a new order predicated on the disappearance of the damaged object, as in razing; or the order of predictable disorder implicit in the inevitable decay of the object as Ruin.

Woods is hardly alone in noting and making use of the abundant possibilities that emerge from within wounds, wreckage, and other states of brokenness or disrepair. Indeed such exploitation of brokenness is not uncommon.

The Axis Dance Company is an Oakland, California-based troupe made up of both able-bodied and disabled dancers. Choreographers working with the company know they can't build their dances around the usual repertoire of movements. There are many ordinary gestures, such as a kick, that a significant portion of the performers simply cannot execute. But what from one perspective appear to be limitations – the restricted range of movements of dancers in wheel-chairs (some of whom may not be able to move any part of their lower bodies), and the effects of that on the choreographic possibilities for them alone or with their able-bodied partners — from another perspective provide the occasion for the creation of dances that explore and build upon gestures and movements that under more ordinary circumstances might not have been imagined. What are called disabled bodies are of course in crucial ways different from damaged buildings or crumbling highways – for though these dancers' bodies certainly deviate from a certain functional norm, they are, given that difference, in fine repair. Still, they are, by prevailing standards in the dance world, impaired bodies. But, like Woods, the Axis dancers, and the choreographers that work with them, do not rush from the "damaged" site: the dances are not predicated on a desire to return the disabled dancers to an earlier more "undamaged" state (or rather, given the impossibility of doing so, on a regret that such restoration is out of the question); nor are the dances created for a new company built around the eventual dismissal and disappearance of the wheel-chair bound dancers; nor do the dances portray the human body as a site of Ruin, of inevitable, intriguingly sorrowful decay.

A related though clearly distinct exploration of the generative possibilities available in brokenness can be found in a strand of Japanese aesthetics according to which there is more beauty in objects that are imperfect or damaged than in those that are flawless and unbroken. A cracked or broken teapot is a welcome and appropriate part of a tea ceremony, among the functions of which is to underwrite the virtues of simplicity and humility. Worn tea utensils bear witness to a history of being handled, and suggest a company of users whom one now joins.

The parallels between what Y. Saito has referred to as this "aesthetics of imperfection and insufficiency"[12] and the attitude Woods describes are strong. There is no attempt to restore the teapot to an earlier, purer state. A broken pot might be repaired, but in such a way as to highlight the fissures and cracks rather than hide or obscure that part of the pot's history. Nor is the damaged pot simply disposed of. And its continued use marks it as different from a Ruin, part of the appeal of which is its having ceased to function in the way it once did. Worn or broken tea utensils open up possibilities that would not have occurred to us when handling the object in perfect or good repair – not only ways of symbolizing certain virtues, or historical continuities, but ways of being beautiful. The gold or other bright solder marking the fault line along which the bowl cracked open has a poignant beauty not to be found in the contours or the texture of an unblemished or flawless object.

Just as some of Woods's critics worry that his celebration of the abundance to be found in brokenness amounts to an implicit endorsement of war and other forces of destruction, so Saito is uneasy with some of the implications of an aesthetics that exploits, even celebrates, brokenness. For such an aesthetics appears to leave unquestioned the difference between people who can't help but be surrounded by broken and chipped household items and those who can choose to do so, to leave unchallenged the entrenched economic and political conditions that create and sustain such differences. Do Saito's worries suggest that, in pointing to the ways states disrepair provide a fecund source of possibilities undreamt of when contemplating the perfect or the undamaged, Woods in effect is underwriting the forces of destruction, damage and decay?

No doubt there is a way in which Woods is counting on brokenness being able to offer a certain kind of consolation, or at least a fruitful provocation. For though he cautions against what he takes to be the all too handy, short-sighted, and often insidious relief promised by projects of restoring buildings, razing them, or bestowing ruin-status on them, he wants us to see, in the rips and tears and cracks and fissures of damaged structures, abundant sources of unpredictable regeneration. The work of L. Langer suggests that however appropriate this search for new growth from "mutant tissue" may be in the case of inanimate objects such as buildings, it may sometimes be inappropriate in the case of human beings.

In a series of books over the last several decades, Langer has urged his readers to pay close and careful attention to the oral testimonies of survivors of the Holocaust. Those observers expecting – or hoping – to learn that there was something redeeming or salvageable or useful in the experiences of the survivors will fail to understand the message that, according to Langer, is unmistakable to those who allow themselves to listen: there is not a shred of consolation to be found in such experiences. Indeed such experiences are defined in part by their resistance to being understood redemptively, as occasions for heroism, or the triumph of the human spirit against impossible odds, or the strength of inviolable bounds among family members or friends, or the survival of human dignity under conditions designed to rip away its foundations. In Langer's *Holocaust Testimonies: The Ruins of Memory*,[13] we are vividly reminded of Woods's worries that damaged sites might be restored to an earlier condition, or razed to make room for a new structure, or left to be relished as ruins, but only because these options are rendered irrelevant in the case of Holocaust survivors: The pre-Holocaust condition of the psyches of survivors cannot be restored. But neither can the memories be left behind, deported, excised. The "ruins of memory" referred to in the subtitle of Langer's book are not, like archaeological ruins, sites to be visited for the bittersweet reminders of human mortality or human hubris, for the exquisite pleasures of decay they provide. Rather, the unforgettable, indigestible, still relentlessly fragmenting experiences are pits of horror and despair from which survivors long to escape.

> "You're not supposed to see this; it doesn't go with life. It doesn't go with life. These people come back, and you realize, they're all broken, they're all broken. Broken. Broken."[14]

In such brokenness there is no "mutant tissue" out of which heretofore unimagined possibilities of thinking and living might emerge: such brokenness "doesn't go with life." As Langer makes clear, this doesn't mean that survivors aren't able to get on with "normal" life. But the lives of survivors have no fecund connection to the "ruins of memory" around which they necessarily must be built, nothing like the fertile evolutionary relation that Woods's architectural structures have to bombed-out buildings, in whose violent tears can be found adumbrations of heretofore unimagined designs. Commenting on "Landscapes of Jewish Experience," a series of paintings by S.l Bak, Langer in fact raises a Woods-like question: "Can life be repaired or replenished by animating the erosions of death?"[15] Not really, he concludes; what these somber paintings, like the oral testimonies, teach us is to be on guard against hopes that in the wreckage of human lives we might find untapped sources of new ways of being.

> "Our moment is not one of fatalistic despair; faces turned toward the past, we do not seek to make whole what has been smashed, but to move athwart the storm into a future in which the debris is more than just a residue; it holds the alternative."[16]

Humans have carved out a rich variety of responses to the fact that we live amidst damage and disrepair. Familiar ways of dealing with such apparent disorder include trying to restore the damaged person or object or site to an earlier state; razing or erasing the disorder and building an altogether new order; allowing the disorder to play itself out into Ruin. Woods's work suggests another approach, one that appears to seek for the evolutionary potential in damage, that looks for, indeed counts on, an ultimately beneficial interaction between the damaged object and those who must decide what to do with them. The possibilities which Woods insists are opened up for architects who have to figure out what to do with bombed out buildings and decaying urban infrastructures appear to be not unlike the possibilities opened up for choreographers who make dances that include people in wheelchairs, or for those who find more beauty in the flawed than in the flawless. Like Woods, they want to face and explore the damage, the brokenness, not find ways (if any are possible) of undoing it. At the same time, we are aware of the moral and political dangers entailed in facile celebration of the bright possibilities to be found in rubble or brokenness. Langer is hardly alone in pointing out that some damage is so thorough, so crushing, so final, as to offer no inkling of rewarding possibilities; others have joined him in insisting that attention to strengthening our capacity to find new beginnings in the "mutant tissue" found in brokenness should surely be matched by attention to our capacity to prevent the destruction which is its precursor and ground. Langer implicitly invites us to reflect on what might remain out of sight when focusing as we have above on some striking parallels between responses to damaged buildings and responses to damaged humans: for example, we tend to be more cautious (or anyway think we should be, even if we aren't) about the way we treat damaged humans than the way we treat damaged buildings. And indeed there is nothing in Woods to suggest that he thinks we ought to treat humans maimed by war the same way he treats buildings crushed in battle or toppled by earthquakes. We are pretty clear about the metaphysical and moral difference between inanimate objects and human beings and the implications of such difference for the way we think about how to respond to damage done to them.

Debates over the appropriateness of restoring ecosystems – debates that this essay does not join but for which it hopes to provide a larger context — suggest that we are not so clear about their metaphysical and moral status. On the one hand, to the extent to which we are confident that ecosystems are importantly different from inanimate, human-made structures, an attitude toward damaged ecosystems parallel to Woods's approach to ravaged buildings may seem cavalier, even dangerous (if for example it ignores or disregards concerns about the toxicity of a given site). At the same time, the extent to which it makes no sense to imagine us forgiving "nature" for the harms it causes us, or "nature" forgiving us for the immense damage we inflict upon it, is one reminder that however much we are part of "nature," ecosystems are importantly different from humans. Such debates will continue not only because of unresolved questions about how ecosystems are and are not like humans, are and are not like artifacts that humans create, and the moral implications of such similarities and differences. They also will be fueled by the fact that the reparative and restorative projects of humans involve what appear to be contradictory attitudes towards that which is being repaired or restored.[17]

Repair is in some respects conservative: it makes it possible for what has existed in the past to continue in some fashion or other into and beyond the present. Repair gets the car running again, the marriage back on its legs, the community able to resume civil functioning. But in order to bring about this link to the past, in order to undo the damage, repair has to do something about those breakdowns, ruptures, and collapses. In this sense repair is interventionist. Indeed, curators of historical buildings have been taught by authorities in the field to think of options before them in terms of degrees of intervention:

> "We can ... classify levels of intervention according to a scale of increasing radicality, thus: (1) preservation; (2) restoration; (3) conservation and consolidation; (4) reconstitution; (5) adaptive reuse; (6) reconstruction; (7) replication."(18)

In bowing deeply toward the past, allowing a previous condition to dictate what needs to be done, repair requires the kind of humility which according to L. Mumford is a hard-earned achievement of *H. sapiens*: ...in dealing with the forces of nature, man's animism got him nowhere. He might attribute willful mischief to a pot that leaked or to a basket that came apart when it was filled...but he could not come to terms with these recalcitrant objects by any amount of sympathetic communication. Eventually, he would have to overcome his anger or indignation sufficiently to patch the leak or reweave the badly woven osier, if he wants to make it perform its function. That humility before the object, that respect for function, were essential both to man's intellectual and his emotional development.[19]

And yet repair is also presumptuous, in its insistence that a given point in the history of something, or a given condition of something, is more important than any other point or condition; for while repair in one sense honors the past, by paying homage to an earlier moment, in another sense it erases the past, by undoing much of what in the meantime has happened. Hence the conflicts, for example, between preservers of historical sites, and those who would repair and restore them, over what moment of history ought to be the community's focus: should the crumbling gas chambers at Birkenau "be restored, somewhat restored or be allowed to fade into oblivion"?[20]

To repair, then, is to enact a complicated attitude toward the past and the pre-existent: repair is conservative, but also interventionist; humble, but also presumptuous; it honors some moments in the past while erasing others. Moreover, while in its service to the past and the pre-existent we find reasons to distinguish repairing something from creating it or replacing it, and in the conservative commitment of repair to continuity we can note its difference from destruction, the assumption that repair occupies a sealed-off place distinct from creation and destruction tends to obscure (a) the creativity that much repair requires, and the repairs and revisions that most creation entails; and (b) the fact that by its very nature repair involves destruction – the destruction of brokenness, of the state of disrepair.

Precisely because repair and restoration are attempts to undo damage, to make up for destruction, this latter feature of repair, the specific kind of destruction inherent in it, may easily slip from view. After all, what intentional or unintentional destruction accomplishes is the end of something, its demise, its irreparability. If repair is about trying to preserve some kind of continuity with the past, keeping some aspect of it alive, destruction is about producing discontinuity with the past, trying to make sure the past is past, that it's over and done with. That the reparative impulse in human beings works against the destructive forces of nature is precisely why *H. reparans* is *persona non grata* at the site of beloved ruins. But still, repair is hedged round with anxiety that the very process by which something is repaired will destroy it – and not just if the repair is botched in some way. Even the most successful repair job destroys brokenness – in fact, the more successful it is, the more the state of brokenness has been vanquished. Indeed that is the hope and the consolation of repair: that undesirable states of brokenness can themselves be broken. For states of disrepair are as subject to damage or destruction as are states of repair. A good repair job amounts to the creative destruction of brokenness. If irreparability is a state of brokenness that cannot be destroyed, reparability is a state of brokenness that can. It's because he doesn't want to destroy brokenness that Woods wants us to think about a way of responding to damage that involves neither restoration, razing, nor charming ruination. What might it tell us about ecosystems that we can, or that we cannot, imagine that responding to damage to them in a Woodsian way would be possible or desirable?[21]

PART III.

Design

Chapter 8

Ecological Art: Visible, Visionary, Practical and Restorative

Jill Brown

Artists have been participating in ecological restoration projects since the 1970s, and yet their work remains little recognized by the larger community of restorationists. My intention in this chapter is to make a case for the essential incorporation of artists, as experts in their own field, into land and water reclamation projects. I hope that the following discussion of "ecological" artists and some of the artworks they have created, will contribute to a greater familiarity with their work and a wider awareness and appreciation for the essential role they can play in our common struggle to improve life on this planet.

Firstly, it might be helpful to distinguish the word "ecological" from "environmental." "Environmental art" is an umbrella term, referring broadly to any art done with reference to the environment. A landscape painting, an assemblage of recycled materials, or a work such as Christo's *Running Fence,* located out-of-doors, may all be called environmental. While the notion of environmental art immediately conjures to many of us images of the land, or "nature," the environment includes not only that which is physically around us, natural or not, but also the entire culture of which we are a part. Therefore, a performance piece protesting an environmental crisis, for example, would fit into this definition. It soon becomes apparent that the general category of environmental art can usefully be broken down into a long list of subcategories, "ecological" being one of them.

The "earthworks" of the sixties, huge projects that involved sculpting the earth itself, were not ecological art projects. Two well-known examples of this work are R. Smithson's *Spiral Jetty*, 6,650 tons of rocks and earth bulldozed into the shape of a giant spiral at the edge of the Great Salt Lake in Utah, and M. Heizer's *Double Negative*, twin gouges in the landscape in Nevada which create a straight line on opposite sides of a small canyon.

It was precisely in order to distance themselves from the latter that Helen Meyer Harrison and Newton Harrison, probably the earliest and best-known of the ecological artists, began referring to their art as "ecological." They agreed that while earthworks such as those of Smithson and Heizer might be "marvelous to behold... [they were] "in the main environmentally destructive." What the Harrisons were seeking to do with their art was land reclamation.[1]

The precise definition of "ecological art" is still open to question. In the fall of 1992, B. Matilsky, curator of the Queens Museum of Art, organized an exhibition she called "Fragile Ecologies." In the catalogue to the show, Matilsky suggests that while she cannot offer a precise definition of ecological art, it is possible to identify certain projects by a similarity of intent.[2] There are other commonalities as well.

An ecological art project takes into account its position within an entire ecosystem. Its aim is not merely to beautify a degraded area but to actually restore or reclaim, or in some cases, prevent further destruction of the environment. In so doing, it may also create a variety of educational opportunities—it may serve as a reminder of the history of the place; it may draw people into closer contact with the natural environment of the area; and it may work as a model for other ways to reestablish harmony among a portion of the earth and its atmosphere and the life that exists there. A typical ecological art project is accomplished outside museum walls, within urban environments or in the countryside; it may relate to a variety of communities. Media that are used range widely from the earth itself to human-made materials. Due to the breadth and complexity of the projects, they are most often not realized by one person, an artist alone, but rather, generally involve collaborations among specialists from several disciplines. It therefore follows that none of these artworks is accomplished in short order; ecological artists all share the attribute of patience. As veteran ecological artist P. Johanson once remarked, referring to a project that has been in the works since 1987, "Like most of my projects, these things don't happen overnight. There's at least a requisite ten years of grief, hostility and permits and meetings and problems and budget cuts and budget restorations and more budget cuts . . . But I hang in!"[3] Ecological art projects are multi-layered, incorporating metaphor and narrative, beauty of form and design, as well as a strong sense of purpose. Finished "products" are functional, in more than one way, and as such provide a challenge to contemporary notions about art. Ecological artworks are essentially sculptural. Therefore, it is not surprising that they are generally created by artists who began as sculptors—an ability to think three-dimensionally helps—and by artists who, at some point in their professional career, came to the realization that objects produced in their studio, however beautiful or thought-provoking or imaginative, did not contribute much in a practical way, toward helping life on a planet in crisis.

L. Hull is a case in point. Giving up on the fruitless gallery scene, Hull moved her art outdoors, into the deserts of Wyoming and Utah, and set about seeing what she could do with her art to help non-human creatures survive out there. There are more wild animals than people in Wyoming, Hull figured, so she might as well use her art to serve them. The result of that move was her first ecological art project, achieved, most unusually, all by herself.

Understanding the crucial difference even a tiny bit of moisture can make to living things in the desert, it occurred to Hull that she could create unobtrusive catch basins for rainfall and snowmelt. She thought about the kind of design that might both serve that function and fit harmoniously into the landscape. Petroglyphs inscribed by indigenous ancestors long ago in the rocks and on the walls of caves and cliffs became the inspiration for her "Desert Hydroglyphs"—graceful, calligraphic stone etchings in rocks, suggestive of delicate hoof marks, such as the one she calls *Scatter* or symbols of the elements, such as *Flowing Water Moon*. Hull did her carving with hand tools, taking care to make her marks at a distance from existing ancient artworks. There are no maps to these tiny, magical watering holes. Whether anyone, that is, any human being, would ever see these offerings was never an issue for Hull; the creatures that live in the desert know how to find them.

In the 1960s, when she was an art student at Bennington College, P. Johanson articulated a vision of "the world as a work of art" that remains at the core of her thinking to this day.[4] Perhaps the best known of her many extraordinary projects is the restoration of *Fair Park Lagoon*, now called *Leonhardt*, in Dallas, Texas. Having seen Johanson's 1978 "Plant Drawings for Projects," a collection of visionary plans for ecologically sound parks and gardens based on graceful forms abstracted from nature, H. Parker, then director of the Dallas Art Museum, decided that Johanson would be just the person to come up with a design to reclaim the environmentally degraded body of water marring the landscape in front of the Museum. That was in 1981. Five years later, the new *Lagoon* was dedicated.

Envisioning the "old mud hole," as Parker called it, transformed into a work of art, Johanson set about researching the history of the land as it had once been, before its shores had begun to erode, the water had turned murky, and the plants and animals had died. By the time Johanson arrived on the scene, the public's association with the water was limited to the knowledge that children sometimes fell in there and drowned. The lagoon was an obstruction to be avoided; it was an inconvenience, taking up the length of five city blocks that had to be circumnavigated in order to get from one side of it to another. With the help of professionals from the nearby Dallas Museum of Natural History, Johanson learned both the causes of the demise of the lagoon, one of them being the inordinate proliferation of algae due to local fertilizer use, and what might remedy the situation. She then designed a sculpture to benefit everyone—the lagoon, a variety of plants and animals, and the people of the city of Dallas.

Two common plants, a Texas fern, *Pteris multifida,* and the delta duck-potato, *Saggitaria platyphylla,* provided the primary design elements for her sculpture. At the northern end of the lagoon where the shore had been eroding at the rate of eight inches per year, she created, out of gunite, (a mixture of cement and sand), an open-work piece suggesting a magnified version of the duck-potato with its tangled roots and slender, swirling tendrils. Part of the sculpture was arranged to protect the shoreline—a leaf and stem running along the shore act as a bulwark against erosion at the head of the island. Some segments were strategically placed to further diminish wave action as well as to provide paths for visitors. Leaf forms that emerge at a distance from human walkways serve as small islands of refuge for animals. Narrow stems of the plant rise up out of the water creating convenient perches for birds. The spaces between the tangled roots are microhabitats for a variety of plants, fish, amphibians, and birds.

An enlarged version of the gently flowing shape of the Texas fern provides the structure for the southern end of the lagoon. Each leaflet emanating from a five-foot wide spine either arches over or seems to float on the surface of the water. The longer spans of archway serve as observation bridges for people while points at which the leaflets crisscross create small basins for flowers such as water lilies and irises and little ponds where one can observe the fish close-up. Aquatic life is encouraged by shallow-rooting plants such as bulrushes and wild rice, placed along the edge of the lagoon. These not only serve as food supply and nesting sites for tiny creatures but further safeguard the stability of the banks.

Johanson's patience and the passage of time have worked to good effect at *Leonhardt Lagoon* in Dallas. Cypress trees planted around the pond now form a shady canopy over the entire sculpture. Various fish reintroduced to the pond thrive, while other forms of life, such as flocks of wild birds, have found their way there, attracted by the rich possibilities for sustenance and protection: "Today the lagoon teems with life," writes W. R. Davis, Assistant Director of the Dallas Museum of Natural History. "Those who understand the intricacies of a functioning ecosystem find particular satisfaction here. A kingfisher visiting for the first time in decades signals that the water is clear enough for this master fisherman to spot minnows swimming beneath the surface. A pair of least bitterns, secretive inhabitants of the vegetative shoreline, moved in the first year and has built a nest and raised a family each of the past five years. Ducks and turtles sun themselves . . ."[5]

Johanson does not have a specific agenda for the people. They are simply drawn into an experience with nature: "I'm not interested in telling people what they should think. But I do try to create situations where people can begin to understand, through their own experience, how the world works."[6] *Leonhardt Lagoon* is very visual; the paths are a bright terra cotta color, rather "big and flashy . . . [but] once you run over to it and stand on it, it's gone. You're there seeing a dragonfly, a shrimp go by. It's all micro-habitat. Nobody looks at the sculpture anymore. It's not about the sculpture and it never was." Her sculpture, she says, is a "come-on . . . like a Venus Flytrap."

Passersby are attracted by the unusual, gay-looking shapes and colors and soon find themselves entering the landscape and exploring the life there. Children run along the paths, peering out over bridges, stopping to bend down over the water to examine the snails, clams, sponges, shrimp, fish, turtles, frogs, and waterfowl that they can see up close living together.

It is this vast array of life, an entire ecosystem that keeps the lagoon healthy. "Johanson's work," says art historian L. Lippard, "(and that of others, notably . . . Alan Sonfist . . . Helen Mayer and Newton Harrison . . .) . . . collaborates rather than competes with nature. (This is a fine line, and reclamation may be a better name for the enterprise—reclamation not only of natural resources but of the function of art.)"[7]

Education is often a key component of ecological artwork. In New York City, ecological artist Sonfist's *Time Landscape* serves as a living memorial, not to a person or an event, but to the land as it once was, in pre-Colonial times.

When, in 1978, after a gestation period of more than a decade, it was finally dedicated, the community responded enthusiastically; the *Village Voice* declared: "That it [is] filled with the types of trees and vegetation that existed here before the area was colonized 300 years ago is a great idea, especially to show those of our number who seldom see or hear a forest that New York was not always concrete and steel."[8]

Time Landscape: Greenwich Village, New York is located on a 9,000 square foot lot in La Guardia Place at the corner of Houston and Bleeker Streets that, when Sonfist first set his sights on it, was a contemporary ruin, the unsightly, potentially dangerous remains of an abandoned tenement building overrun by weeds and trash, broken glass and steel. In 1965, rather than jumping at the artist's plan to clean the space up and transform it into a place of beauty, the National Endowment for the Arts turned him down. After all, Sonfist was proposing a public sculpture made out of living things, at a time when giant constructions of cor-ten steel were defining the current notion of public art. Actually, says Sonfist, "it was [the] artists on the panel who opposed it—one called and said how embarrassed he was for me. . . at that time, planting—which I thought was an absolute necessity—was not considered part of a so-called art vocabulary."[9]

Time Landscape is a recreation of an indigenous forest, designed as a living artwork demonstrating ongoing cycles of growth. After extensive research at the New York Public Library and the New York Botanical Gardens, Sonfist decided to organize his landscape into three parts: at one end there is a mature native oak forest, in the middle, a pioneer forest of gray birch, red cedars, and flowering shrubs, and at the other end the woods taper down to an area of low growth of grasses and herbaceous plants. This provides an opportunity to New Yorkers to witness the unfolding of the natural cycle of the seasons, an experience otherwise available only to city folk in Central Park, (an extraordinary landscape in its own right, but with a different agenda).

Time Landscape in the Village offers not only a history lesson but makes a statement about the need to commemorate nature. Sonfist points out that, just like people, nature has a history that deserves to be monumentalized and treasured. "We honor our human heritage, but we have obliterated our natural heritage in the cities. We have to honor both . . . People must understand that the city is not just concrete, and when they look at a landscaped park, they should know that it is not indigenous to the original landscape."[10]

Small mammals have now moved in and migrating birds visit Manhattan's *Time Landscape* every year. Art historian Lucy Lippard, who lives near the Greenwich Village project, says: "I . . . can vouch that it's not one of those unreal projects that has forgotten death. In winter the *Time Landscape* is a tangle of brush, its beauty ravaged and hidden. In spring you watch it awakening, and in summer it's green and lush . . ."[11] The City Parks Department has profited from Sonfist's research into indigenous specimens. Based on his example, it has added a number of species to its list of "approved" trees, that is, trees residents are permitted to plant. Nowadays, *Time Landscape* is a regular stop on tourists' agenda, listed in the guide books as an official city landmark. It has become a vital part of the fabric of the community. As a result of the success of this project, other *Time Landscapes,* a name Sonfist has copyrighted, have since popped up all over the map.

The work of ecological artists such as Sonfist is becoming more critical as the survival of our planet comes more into question. According to Sonfist, saving the planet, which is really the same thing as saving ourselves, is the most important issue today in which anyone can be involved.[12]

H. and N. Harrison have long shared Sonfist's world view. In the late 1960s, the Harrisons decided that the subject of their artwork would henceforth be the issue of survival.[13] They believe that art, with its special powers of communication, can set up what they call an "aesthetics of survival."[14] In their opinion, artists are in a unique position to effect change in the world. Art, they reason, is the discipline most closely related to metaphor. Most people today turn to the sciences for all the answers. But, say the Harrisons, the sciences "are overly concerned with experiment, testing, measuring and proof...many of their metaphors are limited...The idea that technology is able to buy us out of our problems is an illusion."[15] Faulty metaphors are currently driving our culture. We need artists who can creatively envision positive ways in which the culture can redirect its self-destructive course.

The Harrisons begin all their ecological art projects by trying to uncover the belief systems that first bring a place to a state of crisis. They talk with everyone they can: scientists, politicians, city planners, engineers, and townspeople. By encouraging conversation about the problem, "telling the story" of the place, via installations and readings of accompanying prose poems, in museums, as well as in more prosaic settings such as town halls and school conference rooms, the Harrisons bring about a shift in the way a community thinks about its environment.

The Harrisons agree with bioregionalists who suggest that the world would be better off if political boundaries were drawn along ecological, rather than historical lines. In 1988, the Harrisons proposed a plan, *Breathing Space for the Sava River,* to save one of the last large habitats for the black stork and the sea eagle. Their vision included, not just a nature reserve, as had been suggested by the

Croatian Institute for Nature Preservation, but rather, a whole natural corridor that would run the entire length of the Sava River.

They consulted with botanist H. Em and ornithologist M. Schneider-Jacoby. They assessed the dangers surrounding the floodplain—runoff from a fertilizer plant along the river, toxins excreted by a paper mill and a coal mill, cooling water from a nuclear power plant and other industries. They suggested the creation of wetlands to help purify the water where toxins were entering the river, particular plantings to serve as a natural root-zone purification system, and the encouragement of organic farming that would not rely on chemical fertilizers. They further proposed that water that was being used for cooling purposes by the nuclear power plant be recycled into holding ponds for raising warm-water fish.

Part of the Harrisons' proposal took the form of photographs that documented the current state of the river, its difficulties and its riches, all the way from its sources in the Julian Alps to Belgrade where it empties into the Danube River. The artists' poetic discourse, a report on their conversation about the Sava with Dr. Schneider-Jacoby, was another component of the artwork, which was exhibited first in Germany and the former Yugoslavia and then, in venues all over the world. The Harrisons' artistic approach to solving environmental problems was appealing to Schneider-Jacoby; he appreciated "the picture [they presented] of changing the shape of catastrophe into a shape of opportunity, as the space between ... dikes becomes a corridor ... [connecting] ... habitats."[16]

After seeing their exhibition, the Croatian Institute for Nature Preservation approved the plans, the Croatian Water Department agreed, and the World Bank proposed to provide the funds to pay for the cleanup of the river. Sadly, due to the Balkan conflict, the project was put on hold in July of 1991. Nonetheless, the Harrisons achieved their goal of creating a change in the "conversational drift;" the conversation has, it seems, developed a life of its own and the work the Harrisons began has proceeded without them, albeit in a somewhat different form.

Local authorities and non-governmental groups in five countries (Austria, Slovenia, Hungary, Croatia, and Yugoslavia) have agreed to a plan designed by Schneider-Jocoby, in cooperation with The European Nature Heritage Fund (Euronature), a non-profit organization established in 1987 for the purpose of nature conservation, to create a two thousand square kilometer nature corridor along the two rivers that protect the flow of water into the Danube. Already more than 80% of the corridor is protected. The ramifications of this accomplishment are monumental, positively affecting the water all the way into the estuary where the Danube flows out into the Black Sea.

A unique characteristic of ecological artworks is the element of metaphor that distinguishes it from equally functional but more prosaic land reclamation projects. In 1990, New York artist J. Pinto and landscape architect S. Martino won a competition to design a gateway to *Papago Park* at the boundary that divides Phoenix and Scottsdale, Arizona. The plan that captured the imagination of both the Phoenix Arts Commission and the Scottsdale Cultural Council went well beyond their original hopes.

Pinto and Martino had studied the history of the area and were shocked to discover that *Papago Park* had once been a national monument—*Papago Saguaro.* Due to ecological degradation of the area that resulted in, among other things, the virtual eradication of the saguaro plant after which it had been named, the area had been demoted in status from "monument" to mere "park." Looking further back in time, the artists learned that the land had been home to at least seven different civilizations. The Hohokam, thought to be the original people in the region, had once devised a canal system of irrigation that enabled the desert dwellers to collect rainwater over an area of 800 miles, making it possible for them to harvest crops and essential drinking water. Subsequent cultures sustained themselves using the same system, in many cases taking advantage of the actual structures originally put in place by the Hohokam. Today's Pima people believe themselves to be the descendants of the place's original inhabitants.

So-called "improvements" at Papago Park, such as a golf course and other recreational facilities, had resulted in a dramatic loss of biodiversity to the desert. First, the cholla cactus were removed in the 1930s and then, the bursage cactus in the 1940s. These "mother" plants normally provide a sheltering environment for small seedlings and insects that form the beginning of the food chain in the desert. The decline in food sources led to the decline of other life forms, such as the fox. Once the foxes were decimated, the rabbit population swelled and the cacti, such as the saguaro, soon disappeared, the rabbits having nothing else to eat.

The intention of Pinto and Martino's project was to heal, on as many levels as possible. Their starting point was the symbol of the tree. It would serve as a metaphor for life and regeneration and would also form the physical design of the irrigation system that would begin to restore life to the desert. Locally, it has become known as *The Tree of Life.*

The "tree" consists primarily of an irrigation system constructed out of recycled, locally quarried fieldstone in the shape of a tree extended horizontally across the land. The 240-foot trunk of the tree is the main agent of water collection. From the trunk, branches curve out gracefully in seven directions. The branches act as water-harvesting pools, receiving water from the main aqueduct and distributing it to seven desert farming terraces. The "gateway," as the cities of Phoenix and Scottsdale originally called it, was transformed in the artists' minds into a "spiritual opening, celebrating all of the ancient water harvesting processes . . ." The sculpture represents "a symbolic meeting of Nature and spirit . . . the design calls attention to and celebrates the basic life forces of the desert and its timeless methods of survival and regeneration."[17]

The idea to use the image of the tree for their primary design motif grew out of Pinto and Martino's awareness of the universal meanings given the tree by civilizations the world over as well as the particular importance it held for Native Americans of the region who thought of life as entering into the world through a tree-like channel or stem. Pinto and Martino settled upon seven branches for their tree when they learned that seven civilizations had flourished on the land there. "This fact immediately touched me, for the number seven holds great importance for many cultures around the world and is linked with the achievement of higher forms of consciousness and creation itself," says Pinto.[18]

Adding yet another layer of meaning, the tree at *Papago Park* is placed on the land in alignment with the summer solstice, the longest farming day of the year. Five of the seven vertical markers are also aligned with the solstice; two others place the tree in relation to the Native American ruins of nearby Casa Grande and to Squaw Peak, thus emphasizing a sense of place specifically, historically, and within the cosmos.

The *Papago Park* Project has won numerous awards, among them the American Society of Landscape Architects' prestigious "First Honor" award for a collaboration between an artist and a landscape architect. Pinto and Martino like to stress the fact that their work was a collaboration and that neither could say who was responsible for what. "We informed each other and brought our experience and our expertise to the design, and through that process we developed a project that was better than either one of us could have done on our own."[19]

The park was dedicated in June of 1992. Every year the vegetation thickens and more creatures find friendly habitat there. Photographs taken two and three years after the dedication of the sculpture show the gradual regeneration of flora in the area, as indigenous seeds, reintroduced by the artists, have taken hold and been nourished by the tree-shaped irrigation system.

Pinto and Martino's *Tree of Life* has lived up to its mythology. It has become a place where local people meet to celebrate the summer solstice, as well as to commemorate those who are gone; in 1992 and again, in 1993, a candlelight procession was held in memory of people lost to AIDS. "The spectacle was a moving testament to the fact that public ritual is still a requisite in our society and that art has the potential to become central in the ritual functioning of our communities," Pinto says.[20] Where is the "art" in the project? "You really can't point to the 'art,'" says Pinto, "because it is so integrated and so much a function of its site that you can't pull the 'art' out without the entire project collapsing."[21]

Part of the confusion surrounding ecological art derives from the lack of clarity about which discipline contributes what to a given project. If a public work is done on the land and results in some kind of a park, the temptation exists to call it city planning or perhaps, landscape architecture. But these terms, as we have seen, do not describe the piece adequately. In 1984, writer and curator (and participant in the "Brown Fields & Grey Waters" conference) John Beardsley was not yet prepared to deal with the matter of what to call works in the landscape: "With the . . . recent appearance of . . . collaborations between artists and architects or artists and landscape architects . . . questions have arisen as to the identity of . . . works in the landscape as art. This issue is at once too large and too fruitless to tackle."[22] By 1996, he acknowledged the ongoing blurring of lines between disciplines. "Both landscape architects and environmental sculptors are engaged in common pursuit."[23] Others may be as well.

An unusual collaboration between disciplines is exemplified by the work of R. Chaney, a research scientist at the United States Department of Agriculture, and New York artist M. Chin. Chin's interest in creating a sculpture using plants that could detoxify poisoned earth coincided with Chaney's long-standing but largely ignored theory that certain plants could be used to leach heavy metals from industrially contaminated soil. It took the collaboration of an artist and a scientist to realize both men's dreams in a project called *Revival Field.*

Chin had first studied to be a ceramist. He was fascinated by clay, the material itself, its malleability and its susceptibility to transformation. From ceramics he turned to sculpture. While his early work was political, his shift to a focus on environmental issues was in many ways merely a return to his earliest passions. (Chin's latest work would probably be characterized as political again, while one could also make the argument that his work to save the environment never was a deviation from a political orientation.) Although the *Revival Field* project began as a conceptual piece, that is, a great idea, it developed into one of Chin's most physically demanding, hands-on works to date. The materials involved were plants and contaminated earth and both the form and content of the artwork were realized through a series of transformations that took place as the piece developed.

Having come across an article that mentioned the possibility that certain plants might be used to heal degraded land, Chin began to envision a sculpture that would be a visually appealing structure that moved beyond sympathy for a problem to a revolutionary solution to one. It was years, not surprisingly, from visualization to reality. Chin began by energetically researching any scientific information he could find on the subject. After several months he came across the little known work of Chaney. Chaney had attempted in the early 1980s to convince the United States government to run experiments to harness the ability of plants known as hyperaccumulators to absorb toxic materials through their vascular system. Together the artist and the scientist devised a plan for their sculpture for which funding was provided, eventually, not from the scientific establishment but from the art world.

This was not easily accomplished, however. A grant from the National Endowment for the Arts was first awarded, then abruptly denied. In an unprecedented move, then chairman J. Frohnmayer decided unilaterally to veto an affirmative decision made by both the NEA panel and council. He did not understand how Chin's proposal could be art. Engaging in the patient struggle that typifies public projects, Chin petitioned for a private meeting with Frohnmayer to plead his case. Not only did he succeed in getting the chairman's approval but his action established a precedent whereby any artist may now have the opportunity to communicate directly with the head of the NEA under similar circumstances.

With additional help from the Walker Art Center, and after months of negotiations with public officials, a site was determined and Chin and Chaney were able to begin their first "green remediation" sculpture on a sixty-square-foot section of Pig's Eye landfill in St. Paul, Minnesota.

From the start, Chaney and Chin envisioned their work metaphorically, as a "great circle."[24] Associations with a circle are rich the world over. The circle is often thought of as the perfect shape, a metaphor for the Cosmic Mind or God. The circular mandala occurs as a basic pattern in the religious art of many cultures. It is usually quartered and then even further subdivided. It can be in the form of a square, its four corners representing the four directions, the four corners of the world; in fact, the totality of existence. The spoked wheel, the circle subdivided, is associated with the cosmos, the stars and the moon, as well as star-shaped flowers, some of which, such as wild parsley, are said to have healing powers.[25] The circle can stand for time, as the Native Americans conceive of it, with the past, present and future all existing together. We tell time with circular sundials. The Maya represented time on circular calendars. In geometry as well as in art, the circle represents purity. In Renaissance Italy the circle became the basis for the design of churches. And so on.

The art that Frohnmayer had been looking for was in a number of places. Firstly, there was the design, a circle within a square, the circle crisscrossed by paths dividing it into four equal parts, the central spot thereby assuming the look of a target. The plan was poetic, metaphorical, and practical. Once the plants absorb the toxins from the earth, they can be harvested and recycled. "The aesthetic reality is recreated Nature. The sculpting process starts unseen in the ground below in order to reveal the eventual work, a living, revitalized landscape above."[26] "I thought it was a very poetic principle," said Chin, the most "poetic" part of all being the remediation accomplished by the project.[27]

Practically, the division of space within Chin and Cheney's sculpture established a framework for the scientific experiment that would take place within the artwork. The ground beyond the circle and within the perimeter of the square served as a control area, planted with local grasses. Each quarter of the circular interior space was divided into six parts, planted with zinc and cadmium hyperaccumulators. Each narrow pie-like slice was then further subdivided into four unequal parts radiating outward from the central point of the picture. This created ninety-six separate plots, each of which could be tested for different soil and pH treatments. Six types of plants capable of absorbing zinc and cadmium were planted, among them a hybrid variety of sweet corn, Zea mays and bladder campion, Silene cucabalis. The plot was cultivated by Chin and a number of assistants. It was harvested for three seasons, cut and dried like hay, then burned or "ashed," a process that increases the concentration of the metal to the level of commercial ore. The bounty was sent to Chaney each time for analysis that confirmed the exciting possibilities for using plants to clean up portions of severely contaminated earth.

Since creating the first *Revival Field* together in Minnesota, Chin has entirely turned the work over to Chaney who reports that while he continues work at various test sites around the world, dozens of commercial companies are now offering a variety of services that use plants to remove toxic compounds from the earth based on his and Chin's art. Chaney and associates have developed the commercial technology for removing nickel from soils and they have demonstrated a practical basis for the phytoextraction of cobalt, cadmium and zinc. As far as he and Chin are concerned, anywhere their methods are being used can be called a *Revival Field.* They are thrilled that there are "literally thousands" of scientists who are involved in research on one or another aspect of phytoremediation at this time. Unfortunately, reports Chaney,[28] the market is resistant.

Polluted areas all over the country remain a threat to life and a blight on the landscape because of insufficient will on the part of our government to enforce the cleanup of contaminated areas.

Sometimes, it takes an artist. While it was agronomist Chaney who first wrote about the possibility of using plants to clean up toxic environments, his papers were languishing on the shelves of the United States Department of Agriculture until artist Chin came along and convinced him and the National Endowment for the Arts that the idea was artistic as well as scientifically sound.

Sometimes it takes an artist to convince the world that what seems like a "merely" poetic notion can, in fact, be a very practical and useful plan. Ecological art is ultimately functional; that is, useful, in a variety of ways. "Function is always there when you're talking about ecology," says Johanson, ". . . it isn't just my design but my design fits into the larger landscape; it's ecological in the sense that it allows life to go on."[29]

Five years after the completion of *Leonhardt Lagoon,* the director of the Dallas Museum of Natural History wrote to Johanson from, as he put it, "the point of view of a naturalist interested in the urban environment." "Your arrival was a breath of fresh air. Not only were you sensitive to our concerns, but you also spoke our scientific language and took on the challenge of enhancing the environment, not reluctantly but enthusiastically." Davis was impressed by Johanson's ability to work patiently, for so long, with so many different groups in order realize the project: "Beyond traditional artistic concerns such as form, color and the relationship of the parts, there were the added dimensions of politics, fundraising, engineering, public policy, liability, and plant and animal husbandry. Your ability to deal with this multiplicity of contributing factors enabled a dream to be realized in a bustling urban environment with all the attendant complexities of human behavior...This complex work of art points the way toward a new relationship with nature, a relationship built on harmony rather than conflict, cooperation rather than competition and respectful attention rather than arrogant neglect..."[30]

The Arts and Healing Network, an organization that honors artists whose works contribute to positive change in the world, recently recognized Johanson. "In troubled times like these," says their director, D. Hobson, "it is...heartening to know there are artists like Johanson who are dedicated to using art to make the world a better place."[31]

The Harrisons insist that as "outsiders" invited in to come up with solutions to problems in someone else's territory, they can offer the artist's vision, which involves "metaphor, cross-reference, inclusiveness, and holistic thinking .

. ."[32] The artist's tools are what enable them to devise unique solutions such as planners or engineers never would: "We take up the cultural and political, the esthetic and the ecological, all at once. Generally people in other disciplines only take this up part by part."[33]

Ecological writer C. Manes echoes the view of the Harrisons that some destructive stories have taken hold in our civilization and we need to change them if we are to save the world. Artists have already begun to demonstrate to us that the story of human dominance over nature is a myth, which has for too long justified our careless degradation of the environment on which we, in fact, depend for our survival. "...it's time for our culture, and artists, to change the subject. For the last five hundred years, all we've talked about is Man . . . We need to start talking about this other kingdom."[34]

It was heartening to hear, at the Harvard Design School's conference, the degree to which the notion of mingling fields to accomplish environmental remediation has become accepted. J. Beardsley, one of the few representatives at the conference from the world of art history, emphasized the aesthetic qualities of the restoration project of a gasworks plant in Seattle and spoke of the need for new aesthetic strategies to deal with these complex problems. A. Spirn mentioned the need for beauty as well as function. L. Mozingo made a plea for engaging expression and metaphor: "We need art to let us know that we should pay attention to this place," she stated. When working on an environmental restoration project, clearly, it is no longer a revolutionary idea to "blur the fields" between disciplines, whether landscape architecture and environmental engineering, botany and environmental philosophy, or urban design and plain old art.

However, while a joining of forces was consistently touted at the conference as both desirable and necessary, when it came to presentations of specific projects currently underway around the world, it soon became apparent that the incorporation of artists into these important collaborations is still rare. "Art" was assumed to be a component of something called "art and design," which turned out to refer, in fact, to "design" as in "landscape design," and the category of "artist" was seemingly represented entirely by landscape architects.

Ecological art offers a variety of approaches to both beautifying and healing the planet. Today's ecological artists are sometimes touted as the shamans—those members of society who hold the key to transformation, both personal and public—of the contemporary world.[35] J. Beuys, most often credited with being the first to redefine the role of the artist according to the ancient model of the shaman[36] preached that the artist's greatest value lay not in a particular object that she or he might produce but rather, in her or his vision and imagination. "If the world is going to be saved, "says writer S. Kumar," it is not going to be saved by the politicians or by the industrialists, or by the business people . . . [it is] the artist [who] can still see the relationship of unity between human beings and Nature."[37] "Changing the world is what artists should use their envisioning powers for," says Lippard.[38] Ecological artists are trying to do just that but society needs the vision to include them in the work.

Chapter 9

Art, Nature, Health and Aesthetics in the Restoration of the Post-Industrial Public Realm

Tim Collins

Introduction

For better or worse, we have entered an era of what I would describe as participatory ecology. Ecosystems can no longer face the onslaught of human impacts without some critical support (participation in sustaining well being). Humanity has succeeded in affecting global climate; we have seriously diminished the diversity of life on the planet and recently achieved the ability to manipulate the gene pool. In the quest to control nature and expand material culture, we have discovered limits to that world view. We have fallen headlong into unwanted ownership of natural systems through use and are now faced with significant and troubling responsibilities, as shown in the previous chapter by J. Brown. We must seek new models of perception, understanding and interaction if we are to acknowledge and act upon those responsibilities.

On the following pages, I will look to history, art and aesthetics for concepts and tools that can inform an ecologically and socially engaged art practice. I will begin by locating this discussion in the public realm and describing its relationship to nature. I will then provide a brief history of both the applied and cultural ecologies as a background for my own ideas about radical cultural ecology and its relationship to an emerging area of art practice. I define and describe an informed multidisciplinary eco-art practice that seeks to integrate nature and culture, expanding on the ideas outlined by Brown in Chapter 8. I also define strategic points of engagement for the eco-artist as interface, perception and human values. In the final section, I will explore new ideas in aesthetics. Traditional aesthetics with its focus on fine-art has lost the interest of most practicing artists, its discourse being tedious and circular. The area of environmental aesthetics, however, has the potential to awaken this sleeping dragon, (traditional aesthetics) and put it back into the world in meaningful ways. In conclusion I will extract the concepts and tools that I think are most relevant to those of us that are interested in shaping the attendant metaphors, symbols and narratives that define post-industrial nature.

Locating our Discussion: Public Space and the Commons

I am interested in the systems and ecologies which create the experiences that can be understood as the post-industrial public realm. Post-industrial refers to the shift from carbon-based industrial power and production towards a computer-based economy of information, goods and services that began to happen in the late 1970's. The public realm can be defined at two scales, in relationships between individuals as public space, and as the more encompassing social-political concept of a shared commons. It is easy to think about being "in" a public space. Public space has both its spatial and discursive forms. Public space has a perceptible boundary. We choose to either participate or not participate in public space activities. In contrast, the commons have no real boundary. They are part of the experience of place. The commons are a shared experience that is processed through a social-political lens. Public space is to the commons as skin is to breath in the body. The skin is a clear and perceptible public-place of our body, whereas breath is the body-commons which we all share, as it sustains life. One is an obvious physical artifact and the other a ubiquitous necessity easily overlooked until compromised or removed.

The experience of public space is often framed by place, articulated by landscape, hydrology and/or architecture, and defined by social action. Public space can be found in both terrestrial and aquatic conditions. Another public space is framed and defined by voices of citizens, engaged in discussion about shared aspects of life and the issues of the day. Two or more voices in dialogue create this space, which can be casual (inter-personal) or targeted (civic). It is possible for interested parties to capture both the spatial and discursive forms of public space for private interests. Spaces can be fenced, land purchased and access controlled. Civic discussions can be captured and redefined to reflect powerful interests and to minimize the voices of less powerful interests. Public space is an intimate experience in comparison to the commons. I see the commons as diverse and ubiquitous resources which is generally perceived as too dynamic, too diffuse or too integrated into the fabric of human life to have the kind of value that leads to privatization. Where the experience of public space is intimate, the experience of the commons is expansively diffuse. Despite the collective benefit of the commons they can easily become the target of desire for powerful interests (Hardin, 1968).

There is no greater prize than wealth that is extracted from a ubiquitous, once-public common good redefined as a desirable market resource. For this reason, the meaning, form and function of the commons is constantly shifting. For example, a century ago, rivers were considered unalterable natural commons. In the last century, industrial tools allowed us to re-define their function and manage them as resources for industrial production, minimizing their ecological values. In the coming century, it is inner commons that is the target of speculative desire. Our worldwide genealogical heritage is now the focus of new bio-industrial economies that seek to market organisms and systems previously considered part of our common heritage. The post-industrial era provides significant biological and ecological challenges. First, the external world is affected by a legacy of industrial pollutants that remain in our atmosphere, soils and waters. We are just now beginning to realize that we have been and are affecting nature and the global commons in ways never before thought possible. Secondly, the concept of resource extraction has now descended to the micro-biological level, with market interests scrambling to capture value through mapping, manipulation and patenting of genes. In the sum of these two examples we find a range of meaning, form and function which radically redefines our idea of nature embodied within the concepts of public space and the commons.

Ecological Restoration and Art

The project of ecological restoration (like preservation and conservation before it) requires critical and radical (socially transformative) cultural components as well as pragmatic and rigorous science if it is to succeed. The project of restoration seeks to shift the environmental dialectic from a culture that sees utilitarian value in nature – with preservation as the critical solution to industrial landscape changes; to a culture that sees intrinsic value in nature – with restoration as an essential response to post-industrial legacy pollutants and global impacts. On the pages to follow I will describe the role of art, design and aesthetics in the contemporary project of ecological restoration.

Restoration Ecology: The Cultural Aspects

The emergent area of knowledge known as restoration ecology is a logical response to the post-industrial era. Preservation and conservation emerged in the years around the turn of the 20th century in response to the tools and economies of the industrial era and growth and development in the American West. As shown in previous chapters of this book, restoration ecology is a new way of thinking. It links citizens and experts, as well as cities and wilderness, in a broad program of ecological awareness and action. It is a community of disciplines synthesizing a continuum of diverse cultural practices. On one end lie the arts and humanities, in the middle are the design professions, at the other end, science and engineering. Restoration ecology has been touted as a new relationship to nature, one in which the old reductionist paradigm is reversed. Scientists are charged with re-assembling a working nature from the pieces discovered over the last 200 years, while taking it apart. While the machine metaphor was useful in the disassembly and analysis of nature, it is less useful when reassembling nature. The aesthetic roots of restoration ecology can be found in the urban-nature design projects of Frederick Law Olmsted (particularly the Fens of Boston, 1881) [1]. (See also the Chapter 10 by L. Mozingo that follows.) The roots of its' science can be found in Aldo Leopold's work restoring the lands of the University of Wisconsin-Madison Arboretum in the 1930's (Jordan, 1984).

In Jordan's original document, restoration ecology was interpreted as a mixture of cultural and scientific efforts, "....active as a shaper of the landscape, yet attentive to nature and receptive to its subtlest secrets and most intricate relationships. The restorationist is in this sense like an artist and a scientist, impelled to look closer, drawn into lively curiosity and the most intricate relationships" (Jordan, 1984). After Leopold, Jordan is clear that restoration is about restoring a "whole natural community, not taking nature apart and simplifying it, but putting it back together again, bit by bit, plant by plant", "....the ecologist version of healing. " (Jordan, 1984) Jordan commented on the import of restoring whole communities in this text, but he also recognized the import of restoring (reclaiming) industrial sites, referencing the noted biologist A. Bradshaw's pioneering work on coal mining sites in England. Jordan sees the Arboretum as a research laboratory for work that will be in increasing demand in the future, due to the fact that the industrial revolution has provided humanity with the tools to affect nature on a grand scale.

Restoration ecology attempts to both define and reconstruct nature while staying aware (and respectful) of the complexities of the process, its ethical context and the social potential of its performative aspects. Restoration ecology is an important new area of thinking and acting. It provides us with experience and knowledge that can transform the human relationship to nature.

Art and Ecology

For clarity, I want to describe arts relationship to ecological restoration in divergent yet complicit terms. First it is a fine art activity, with a relationship to the critical and intentionally socially transformative components of the historic avant garde. It is also a design activity, which is about the organization and application of content within a known context with a clarity of intent that produces form. The former is based in a tradition of creative autonomy has more propensity for a critical and radical stance. The latter is based in a tradition of creative response to the needs of a client. Framed through critical knowledge but ultimately is complicit with dominant interest. It is in the relationship between these two ways of working, (and many of us, wear both hats) that the arts serve nature and culture and the project of ecological restoration.

As stated previously restoration ecology has clear intent to change the human relationship to nature. A branch of the arts has coevolved with a similar idea which I will describe below. As summarized by Brown in chapter 8 there is a rich tradition of artists working with the environment in terms of landscape painting, earth works and ecosystems approaches. However, most of this work was created from the philosophical position of an increasingly entropic avant garde, that had severed all relations to the social and political mileu to focus upon free creative expression and the pursuit of formal and contextual innovation. This is why earth art was such a visual and theoretical sensation. The decision by artists to take their formal sensibility out of the gallery, was a radical and transgressive act. At the close of the 1970's the primary intent of all but a few artists was to steer clear of utilitarian social political and environmental issues. The primary role of the artists of that time, was one of unmitigated creative freedom...driven by a quest for innovation, an innovation without social consequence, but that was about to change.

Writing in the 1980's, the art critics L. Lippard author of "Overlay" and

(1983) and S. Gablik, author of "Art After Modernism" (1984) provide us with the theoretical and conceptual impetus to reintegrate art with society and its increasingly troubled relationship to the environment. Their project was to theorize an integration of the individual as a moral, social and ecological being. As socialists and feminists, they shared a critical unease about the artworld and capitalist society, as well as a desire to restore meaningful [2] traditions and transformative practices that embrace the full knowledge of the human condition, (both the masculine and feminine) and reintegrate art, society and the environment. More recently, in the last fifteen years, critics and theorists R. Deucsche In "Evictions"(1994), M. Miles in Art Space an The City" (1997, 2000, 2004), and T. Finkelpearl author of "Dialogues in Public Art" (2000), provide us with a critical baseline and theoretical standard for art informed by the social issues inherent to the discourse and agency that attend planning and development. They have theorized directions that addresses radical (socially transformative) forms of public art that can integrate social and ecological concerns.. Recently the author M. Kwon, writing in "One Site After Another" (2004) has examined the role of the artist in both place based and discourse based creative practice. G. Kester writing in "Conversation Pieces" (2004) has begun the task of theorizing a dialogic aesthetic. My goal here is to present the non-artist reader with an understanding of shifts in the theory and practice of just one sector of the arts, in recent years. The results can be described as a transformative form of art-practice that locates the artist in dialogic relationship with place based communities and ecosystems. The "form" of this work, whether it is a physical product, a plan or a process, has been shaped by multiple hands, and as a result has multiple advocates

Accompanying this critical and theoretical literature, groups of artists are emerging who are committed to a social-ecological process that go beyond the ideas of authorship and creative identity typical of previous generations. Many of these artists work collectively or collaboratively. Some of them pursue lives of art and social/environmental activism following the examples of the German artist J. Beuys[3]. Others are involved in issues of art, science and planning following the examples of artists such as H. and N. Harrison, of California.[4] The current generation of artists can be divided into two groups. First, there are eco-artists who are interested in the integration of nature and culture through concepts and practices that are informed by ecology and natural systems. Secondly there is an art and bio-technology movement defined by an interest in how the new bio-technologies affect questions of humanity, nature and culture. While I am interested in both groups of practitioners (and there is some overlap), I participate socially and intellectually in an international group of eco-artists, scientists and planners, and will focus my discussion there for the remainder of this chapter.

Eco-Art

The name "*eco-art*" is a term used by many, although its definition and intent is variable. Art critic Lippard (1983) defined it as having an "emphasis on social concern, a low profile, and more sensitive attitude toward the ecosystem." Recently others have begun to work through the meaning and intent of this way of working. One of the most consistent thinkers and authors on the subject is R. Wallen a San Diego California practitioner with training in both art and biology, and a former student of the Harrisons. Wallen offers the following definition, and guidelines written with members of the international Eco-art Dialogue group. (http://www.ecoartnetwork.org)[5]. This quote has been edited for brevity.

> "Ecological art, or eco-art to use the abbreviated term, addresses boththe heart and the mind. Ecological art work can help engender an appreciation of the environment, address core values, advocate political action, and broaden intellectual understanding.
>
> Ecological art is much more than a traditional painting, photograph, or sculpture of the natural landscape. While such works may be visually pleasing, they are generally based on awe-inspiring or picturesque, preconceived views of the natural world. Ecological art, in contrast focuses on the system of ecological relationships. These relationships include not only physical and biological pathways but also the cultural, aspects of communities or ecological systems. Much ecological art is motivated by a recognition that current patterns of consumption and resource use are dangerously unsustainable. Instead of focusing on individual gain, ecological art is grounded in an ethos that emphasizes communities and interrelationships.
>
> The focus of a work of art can range from elucidating the complex structure of an ecosystem, responding to a particular issue, interacting with a given locale, or engaging in a restorative or remediative function. Ecological art encompasses both process, i.e. design and planning, and product in the form of a discrete work of art. Eco-art may re-envision, or attempt to heal, aspects of the natural environment that have gone unnoticed or reflect human neglect. The work may challenge the viewer's preconceptions and/or encourage them to change their behavior.

> Ecological art exists within a social context. While certain works may express an individual vision, the intent is to communicate—to inspire caring and respect, stimulate dialogue, and contribute to social transformation." (Wallen· - written with members of the eco-art dialogue; http://communication.ucsd.edu/rwallen/ecoframe.html)

Wallen's text, developed with others – makes it clear that the artists role is based in values and advocacy for an ethical ideal of collective networked relationships that reintegrate the social and the ecological. The product of the eco-artist can be design, planning and/or the manufacture of isolated objects of art. Another text, that might shed some insight on this evolving area of practice has been developed by the art historian and author L. Weintraub working with artist S. Schuckmann. The initial idea presented in the form of a manifesto is described (in edited form) below. I have edited the text for brevity.

> "THE PREMISE Eco artists are distinct from other artists because they sculpt their impulses with full consciousness of the effect of their work on the environment of the planet and the distribution and abundance of organisms that inhabit it.
>
> A. MICRO Eco art, engages the intimacy of home (here) and the immediacy of time (now). It specializes in here and now by valuing indigenous materials, locally generated energy sources, sustainable procedures, and topical themes.
>
> B. MACRO Eco artists are mindful of our universe, galaxy, solar system, atmosphere, hydrosphere, and geosphere. They consider the mutuality and interconnectedness of all forms of life. The scope of ecological science means that it cannot be limited to laboratory procedures. Similarly, eco art is not confined to the studio.
>
> C. MUCKRO is a term invented to honor the middle zone in which people actively engage the complex "muck" of everyday life. "Muck" is dense and murky. It is difficult to comprehend and navigate. It is also a massive ecotone of human potential.
>
> The goal of eco-art is to develop functional awareness of the proximal and distant impact of human behaviors on the living and non-living environment, and the environment's impact on human beings. It considers the

> health of our bodies, our local biomes, and global eco systems." (Weintraub and Schuckmann; http://www.lindaweintraub.com).

Weintraub's and Schuckmann's manifesto extends Wallen's effort in important ways. Where Wallen retains the frame of the ecological system, Weintraub and Schuckmann name the components of the ecosystem, both the "living" and the non-living" extending recognition to other species. This raises an important question of spirit, the very principle of life, that is present when you are alive, and gone when you are dead. It refers to the immaterial intelligence, the aware sentient side of our very being and the increasingly flexible boundary that defines sentience within us, and amongst the creatures around us. This is an essential topic of new moral and ethical analysis in philosophy. While I am sure that both (groups of) authors share an awareness of organisms, couching the environmental context in systems language alone – reinforces the human as dominant species. Not that naming "organisms" is enough to shift the historic nature-culture dialectic to a position of equitable representation, but none the less it is sometimes important to state the obvious. The two texts share a primary interest in sustainability, with Wallen making it clear that she believes that "current patterns of consumption and resource use are dangerously unsustainable." Weintraub's and Schuckmann's approach is more prescriptive (yet idealistic and simplistic) with a guideline to use "indigenous materials and locally generated energy sources." Finally it is important to state for the non-artist, what it is these authors claim that eco-artists do. First they both claim that art provides an awareness or understanding through the design, planning and/or the creation of isolated objects. In addition, Wallen claims that eco-art can "address core values, (and) advocate for political action." Weintraub and Schuckman raise one additional issue, the idea that we (our bodies) are linked to the environment through issues of health. I will return to this at the end of the chapter.

Ultimately, from my point of view, this work is about public realm advocacy. Advocacy for communities, organisms and entities that are not well represented within the traditional dialogue between state and capital. I would argue that the eco-artists role is to develop interface between nature and culture, and act as an agent of change. Like my colleagues above, I believe it is our function to reveal concepts and experiences that might otherwise be overlooked. As a result I would define the practice Ecological Art or Eco-art: as a creative process that results in interface between natural systems and human culture. It recognizes the historic dialectic between nature and culture and works towards healing the human relationship to the natural world and its ecosystems. Where much of the art (specifically the avant garde art) of the past has focused on a critical relationship to culture, eco-art focuses on a critical responsibility for the reintegration of nature and culture. In this, we are not unlike our colleague in restoration ecology. I would argue that the difference is that we are primarily working on restoration at the level of perception, conceptualization, experience and value. While our colleagues in engineering, the natural sciences and social sciences are working on restoration through the renewal of structural systems and interacting networks of nutrients and organisms. The action of our colleagues can result in the restoration of health to complex systems. The actions of eco-artists call into question the cultural relationship to nature. And, at times we use the tools of science to accomplish our goals. I should say that we rely upon the work of eco-philosophers who have been instrumental in clarifying the increasingly complex moral and ethical issues that define the nature and culture relationship at this point in time.

Current Exhibitions and Relevant Texts

While eco-art remains primarily outside of the arts mainstream, non the less various curators, critics and authors have seen the need to address the work through exhibition, catalogs and critical texts. B. Matilsky curated and exhibition "Fragile Ecologies"(1992) and overview of the area of practice, with an accompanying catalog,. She provides an excellent overview of the historic precedents for this work, as well as some of the most important work of the first and second generation of ecological artists. A text edited by B. Oakes, "Sculpting with the Environment" (1995), is unique and quite valuable as a reference in that he asked artists to write about their own work. "Land and Environmental Art," an international survey of both types of artists projects, was edited by J. Kastner, with a survey of writing on the subject by Brian Wallis (1998). The text goes into the first, second and third generations of earth and ecological artists, providing an overview of works and accompanying articles. In (1999) H. Strelow curated "Natural Realities" at the Ludwig Forum in Aachen, Germany, this exhibit was an international overview, which expanded the concept of ecological-art and its range of effort to include the human body as a site of "natural" inquiry. The accompanying exhibition catalog provides cogent arguments for the three areas of the exhibition - the unity of man and nature, artists as natural and cultural scientists and nature in a social context. The first exhibit to attempt to directly address the new ideas and instrumental intent of eco-art occurred at the Contemporary Arts Center in Cincinnati, Ohio. S. Spaid and A. Lipton (2002) co-curated *"Ecovention: Current Art to Transform Ecologies"*i. The accompanying catalog explores the artists role in publicizing issues, re-valuing brownfields, acting upon biodiversity and dealing with urban infrastructure, reclamation and environmental justice. In (2004) a new book, on "Ecological Aesthetics" initiated by H. Prigann, a German ecological artist and edited by Strelow and V. David, provides an excellent overview of the range of work that is occurring today in both Europe and the United States. Another amazing new international resource for those interested in the range of work in this area of practice, an excellent source can be found online (only) at the GreenMuseum.Org, (http://www.greenmuseum.org) This is a project developed and directed by S. Bower.

Art and Radical Political Ecology

I want to think of interface as a common boundary or interconnection between systems, equipment, concepts or human beings. Interface is the art, the physical manifestation of the "relationship between humanity and the natural world." The concept of interface is appropriately open. Its form is undetermined but its intention is explicit: it defines the art of ecology without closing out its options. Perception is the awareness of interface or awareness through interface. Human values are the target or goal of cultural agents. (The active role of agency is assumed under the interdisciplinary model.) Eco-artists manipulate the attendant metaphors, symbols and narratives of the nature/culture interface to shift human perceptions around the dual subjects of their inquiry, research and production - affecting valuation. These are the strategic points of political engagement for the eco-artist - interface, perception and human values.

Theory and Interdiscipline

As an eco-artist with an interest in philosophy and theory, I am interested in form, content and symbols as well as the concepts and theories that inform and sustain the practice. I would argue after the Harrison's and Sonfist, (Auping, 1983) that eco-art is fundamentally interdisciplinary, in that we can not rely on the art world as the only point of engagement and interpretation. Furthermore, the artists involved in this practice can't confine their learning or production to art alone. We must reach out across disciplines to build a platform of knowledge and practice. In the interdisciplinary model, artists find critical social space to expand their practice by moving outside their discipline and its institutionalized relationship to society. In this way, we find opportunities, both intellectual and creative, that we cannot find within our own discipline (which like most other disciplines has turned inward upon itself.).

Interdisciplinary practice breaks the form of discipline specific institutions. It expands the combined disciplines and provides the artist with a new path to social engagement. Inherent in that path is the responsibility for the artist to educate him/herself in several disciplines. In turn, the work needs to be received and evaluated for the totality of its intention.

Dualities and The Philosophies

It is important to understand the philosophies that can inform our actions. The environmental movement can be broadly characterized by a struggle between the oppositional ideas of nature as an autonomous and intrinsically valuable entity unto itself versus nature as both concept and focus of human exploitation for economic value. The social-ecologist M. Bookchin (1974) sets up a simple duality, to help us better understand these ideas: "Ecologism refers to a broad, philosophical, almost spiritual, outlook toward humanity's relationship to the natural world ...Environmentalism [which is] a form of natural engineering seeks to manipulate nature as a mere 'natural resource' with minimal pollution and public outcry." Bookchin's position on ecologism and environmentalism is comparable to the duality of preservation and conservation. But what does this mean for artists? First, ecologism provides artists a pathway into a new area of knowledge. In that broad philosophical/spiritual outlook there is plenty of room for artists to experiment with interface, perception and human values. There is less room for artists in his concept of environmentalism.

Where do we stand in relation to nature? We can broadly situate ourselves in either the wilderness or the garden. (Mitchell, 2000) This simple duality allows us to consider wilderness as the condition of nature without human impact, and garden as the human condition (or city condition) of nature. Our value systems can flow in either direction. If value is centered in the garden, then it is the use of nature that drives our actions. The garden relationship assumes that we are above nature and capable of some charitable (and not so charitable) contributions to nature. If value is centered in the wilderness, then it is the maintenance of that boundary separating humanity from nature that drives our actions. Wilderness (by strict definition) is a condition that can be defended, defined or interpreted but never improved upon by human action. Three philosophies have emerged which inform a continuum of human thought and action in relationship to garden and wilderness ideas: social ecology, (Bookchin 1980, 1982) eco-feminism (Merchant 1980, 1982; Plumwood 1993) and deep ecology (Naess 1989; Sessions 1995). These three ecologies share a common thread — the negative affect of human civilization upon natural systems has instigated the need for various radical communities to seek a path to action.

These three philosophies, with their spatial commitment to city, town, country or wilderness, and their political commitments to humanity, post-dominion humanity and the intrinsic rights of nature itself, provide a broad intellectual foundation for the eco-artist. This is a foundation of human values that project ways to understand and act in relationship to nature. This foundation provides room for a range of practitioners with a shared interest in the roles that art can take in the changing human relationship to natural systems. This foundation can accommodate the artist as witness, advocate or activist, but always as an agent of change in the shifting values of nature and culture.

Aesthetic-Ecologies

Nature has been a fundamental subject of artistic practice and aesthetic inquiry throughout history. Nature has filled the artist with fear, awe and wonder. Only recently has the material product of the artist, artwork become the sole subject of the philosophy of aesthetics. Since the 18th century, the dominant western philosophy of aesthetics concerned itself with the appreciation of things deemed pleasing, or things with the potential to evoke an experience of the sublime. (In minimal opposition: Marxist aesthetics has been more concerned with the social relationships of production.) The operative word here is "things," isolated objects that exist independent of context and those that view them. The concept model is simple: a human appreciator and a thing, framed in a neutral manner, which is then appreciated. The means of appreciation was primarily visual. The objects of consideration were carefully bounded to separate art from daily life. The viewer was expected to be properly (empirically) disinterested in the object of contemplation. These things were then analyzed for beauty paying attention to their unity, regularity, simplicity, proportion, balance, measure and definiteness[6]. Alternatively, works could be analyzed for their relationship to the sublime, the feeling of sublime emerging when a viewer considers an object which sets up a tension between imagination and reason. In the contemplation of the finite object, we find an experience of expansive grandeur, wonder or awe. In this historic model of aesthetics, the world is left to rational utility. These ideas of beauty and wonder are exclusive, properly separated from that world and confined within reductionist laboratories that let us see the work without the corrupting influences of social, political or environmental conflict. The white walls of the museums, the raised stage of the symphony, or the frame of the painting all provide us with a clear understanding of where to look and contemplate objects for their inherent aesthetic value. Modernist asethetics has no value for artists that have embraced the post-studio practices. Ecological artists, informed by earth-art and enabled by the freedoms of post-modern multi-disciplinarity, fit neither the context nor the method for aesthetic analysis. Ecological art relies upon experiences that are enmeshed in complex natural systems. Authorship lies on a fine line between action and concept. Relevant form rarely stands alone. More often form is extracted from the context itself. Complicating things immeasurably, there is a whole social-political element of the work that cannot be ignored. The elite, disinterested root of aesthetic philosophy would seem a long way off from art practice focused upon strategic engagement with interface, perception and human values.

Complicating the Meaning of Nature

In a controversial article with ongoing repercussions, philosopher R. Elliot (2000) claimed that the practice of restoration ecology is nothing more than counterfeit nature, as egregious (and worthless) as a counterfeit of a great painting or sculpture. He declared that wild-nature had an irreplaceable natural quality, as irreplaceable and authentic as a fine-art masterpiece. He further declared that the practice of restoration ecology when applied at a policy level allowed developers and extractive industries to destroy authentic nature.(7) Replacing natural authenticity and intrinsic value with a counterfeit or restored ecosystem calls into question our moral, scientific and creative potential: can we only save or destroy nature? E. Katz, A. Light and C. Foster have all addressed this question of counterfeit nature in different ways. Katz states that "the natural is defined as independent of the actions of humanity", which in turn results in his position that "we do not restore nature, we do not make it whole and healthy again." (2000: 90) In the same edited text, Light answers by granting Katz the claim that it is impossible to restore nature (as Katz has defined it), but he contends that we still have a moral obligation to improve and refine the technological and cultural projects of restoration, restoring what he calls *"the culture of nature."* (2000) In contrast, Foster (2000) comes at the question of authentic nature from an environmental aesthetic position. She suggests that within the United States, the authenticity and trust in restored ecology, geology or nature in any form, will be consistently plagued by a cultural tendency towards hyperreality and the simulation of nature. The author explores the restoration and maintenance of "natural-wonders" at parks and national recreation areas. Four philosophers who carry three views of nature. The first and second accept nature only in its independence of humanity (a notion I strongly disagree with for reasons that will follow), the third seeks a culture of nature, and the fourth points out that, not only do we have to deal with the natural and the restored, we also have hyperreal nature to contend with. Whereas Light calls for a culture of nature, Foster describes a nature of culture which bends the meaning of the former in ways that are only constrained by the imagination.

Most of us have strong feelings about nature. We arrive at these feelings through a range of "natural" experiences and cultural training. Can we know nature without compromising its independence? Environmental restoration challenges this understanding of nature in odd ways. Is it enviro-technical or is it enviro-medical? We can approve of medical intervention for humans, pets and livestock. We even perform wildlife rehabilitation in most of the major cites in the country. We can approve of technical soil remediation, species selection, and ambulatory plant care for desirable flora (such as lawns in Las Vegas). But the idea of restoring nature and usurping its wild integrity generates a passionate defensive position in the most liberal corners. The passion that is elicited to defend disappearing ecosystems, disrupted landscape ecologies and their related organisms against a loss of authenticity truly puzzles me. The tools and economies of the industrial age have left us with an awesome ability to shape, mold and transform nature into the material goods of culture. Natural authenticity is physically compromised by industrial by-products that exist in the air, water and soils. Conceptual authenticity (wildness) is compromised as we discover, name and catalogue the genealogical complexity of nature. What we can not get to physically and conceptually, the global climate change will. Given the inalterable fact that nature has been and will continue to be compromised, do we have an ethical duty to preserve, conserve and restore what we can? If we do, how can aesthetics help us in this expansive project?

Environmental Aesthetics

There are a number of important thinkers in the area of environmental aesthetics. Those working from an environmental psychology point of view, such as J. Appleton, R. Kaplan, S. Kaplan. And those working from an environmental philosophy point of view such as J. Nasar, Foster mentioned earlier, J. Nassauer and M. M. Eaton discussed later are just a few. I will look into the work of two primary voices in the area of environmental philosophy next. They are A. Berleant author of *"The Aesthetics of the Environment"* (1992) and A. Carlson author of *"Aesthetics and the Environment: The Appreciation of Nature, Art and Architecture"* (2000). In a co-edited volume of the *"Journal of Aesthetics and Art Criticism"* (Vol 56, Number 2, Sp 1998), they define environmental aesthetics at face value as "the application of aesthetic concerns to environment." This concept is almost the polar opposite of the traditional aesthetics outlined earlier. First the term environment qualifies aesthetics in important ways. It is inclusive and expansive, opening this philosophy to a range of culture of nature and nature of culture conditions that would not be considered under the exclusive and reductionist methods of more traditional aesthetics. Qualifying aesthetics with environment also raises the idea of application. Once aesthetics accepts the challenge of finding the means and methods of describing aesthetic value in complex and diverse environments, the application of that knowledge is likely to follow. (Whether it will affect the dominance of economic-production value is another question entirely.) Most importantly, however, in the combination of environment and aesthetics, a reconstructive post-modern path is drawn out of what could be described as a reductionist endgame seeking a truth that has decreasing relevance. In environmental aesthetics, the full range of nature-culture manifestations are opened up to multi-sensual perception, emotional and intellectual analysis and social-aesthetic evaluation. What was once simplified in the pursuit of empirical truth has become complicated and complicit with the world once again. The question is, can environmental aesthetic philosophy handle the complex experience of dynamic systems with intellectual tools developed over the last two centuries studying static self-referential objects of fine-art? Can art and aesthetics provide us with more sophisticated tools to conceptualize nature than Elliot's dichotomy of natural authenticity versus restored nature as forgery? In the following pages you will see this struggle for an aesthetic understanding of nature manifest between the two approaches of Berleant and Carlson.

An Aesthetic of Engagement

Berleant is a philosopher and a trained musician interested in both the theory and application of his work. Since 1970, his provocative and bold writing is intended to expand the focus and purview of aesthetic philosophy. In *"The Aesthetics of the Environment"* (1992), Berleant outlines an aesthetics of engagement which seeks ultimate unification of nature and culture, declaring "there is no sanctuary from the inclusiveness of nature."(1992) In this model, Berleant outlines a radical aesthetic theory that casts aside the subject-object[8] relationship for what I would describe as an integrated systems analysis[9] approach to aesthetics. In this theory, nature and humanity are one field. Artifacts as the material product of culture are no longer isolated and the disinterest which has marked two centuries of aesthetic philosophy gives way to passionate engagement with contextual experience. Berleant claims that "The aesthetic is crucial to our very perception of the environment. It entails the form and quality of human experience in general. The environment can be seen as the condition to all such experience, where the aesthetic becomes the qualitative center of our daily lives." (1992) He works to provide an aesthetic paradigm intended to open the world to a "full perceptual vision of aesthetic, moral and political conditions." (1992). His proposal is based on the following three points: 1) the continuity between art and life; 2) the dynamic character of art; and 3) the humanistic functionalism of the aesthetic act. He applies these ideas to the city, working to develop what he calls an aesthetic paradigm for urban ecology. The components of his paradigm, (1992), with examples, are:

- The integration of purpose and design as typified in a sailing ship.
- The integration of fantasy and spectacle, subhuman and human, found in the circus.
- The communion between heaven and earth, sanctuary and steeple found in a cathedral.
- The union between the individual and the celestial, organism and cosmos found in a sunset.

These four components are described as typical dimensions of a city that are overlooked, subsumed or subordinate to utilitarian development. In turn, they are presented as strategic interventions in cities to achieve a "critical measure of urban aesthetic." I am in agreement with Berleant's "aesthetics of engagement" but find the examples limited. It occurs to me that what he has left out is any sense of a critical-social or creative-social approach to art and urban ecology. He has kicked aesthetics into the present but left art and natural science in the past. There is no sense of the artist or restorationist as a strategic cultural agent acting with full awareness to shift the symbols and metaphors of a culture invested in the power of state and capital who are in turn, invested in utilitarian approaches to cities. The historic components presented by Berleant provide us with a historically referential framework for a culture that integrates the aesthetic with the functional. It does not give us the right tools to achieve those goals in contemporary culture. Glorious sailing ships, spectacular circuses, breathtaking cathedrals and cities oriented to the sun emerged in cultures that put primary value on those things. The integration of the subject-object provides us with a new conceptual framework. But, the components of the paradigm are passive and more likely to conform than to transgress. Integration, communion and union are based on relationship. The culture of capital and its utilitarian approach to city building are the dominant economic and political power. Reestablishing humanistic-aesethetic values in a culture of capital will require a strategy that is both cognizant of that power and able to develop strategies to achieve the desired relationships. Artists and aesthetic philosophers who are committed to an aesthetic of engagement are going to have to get realistic about the application of their ideals. This will be a significant challenge, I would add three components to his paradigm to open up that potential:

- The critical relationship between society and art, aesthetics, morality and equity.
- The creative relationship between places and people, need and awareness of limit
- The respect and ethical rights that are shared or denied amongst sentient beings.

A Natural Environmental Aesthetic

Carlson's (2000) work in *"Aesthetics and the Environment: The Appreciation of Nature, Art and Architecture"* is a more deliberate approach to environmental aesthetics. The depth and rigor of his analysis is quite remarkable. This is reflected in his conceptual organization of the issues and models for aesthetic appreciation of nature. He begins by defining the scope of environmental aesthetics in terms of the range of things we are to consider, from pristine nature to human art and cities. He also defines the environmental aesthetic scale from objects to bounded properties and forests. (he does not identify ecosystems or the nature-commons.) He identifies the range of experiences, from mundane to spectacular, and goes on to talk about the complex experiences that can be found in even the most common forms of nature. His stated goal is to create a set of guidelines for aesthetic appreciation that will allow "serious and appropriate interpretations" of nature. Answering the "what" and "how" questions is one of his essential preconditions for genuine aesthetic appreciation. He describes two basic orientations when we attempt to appreciate nature aesthetically. The first he describes as subjectivist or skeptical, whereby the viewer is frustrated by nature's lack of frames, design and designer. (the viewer does not know what or how to appreciate the unframed landscape) His second point is described as objectivist. "In the world at large we as appreciators typically play the role of artist and let the world provide us with something like design..." (2000) If I understand him correctly, within the recognition of pattern, we can then set boundaries which allow us to define the "what" which then provides the question of "how" to appreciate nature. He provides specific ideas about categories or models which can inform the appreciation of nature.[10]

Carlson concludes that the natural environmental model and its close ties with scientific knowledge is the right approach. He sees its roots emanating from a tradition of thinkers like George Bernard Marsh, Henry David Thoreau, John Muir and Aldo Leopold. (I wonder if Berleant would not claim the same roots for his aesthetics of engagement) Qualifying the aesthetic with the scientific adds a cachet of objectivity that he believes is important if aesthetics is to have any impact on practical environmental assessment. He is quite clear in his position, "…appreciation must be centered on and driven by the real nature of the object of appreciation itself. In all such cases what is appropriate is not an imposition of artistic or other inappropriate ideals, but rather dependence on and guidance by means of knowledge, scientific or otherwise, that is relevant given the nature of the thing in question." In this bold statement, Carlson makes his own definitive leap for aesthetic philosophy, distancing it as far away from art as possible. Carlson grounds aesthetics in knowledge which I agree with, but I feel the need to question the standard of science as the only path to knowledge. (This approach is central to the technical aspects of the project of restoration ecology.) Authorizing science and disavowing the sensual, kinesthetic, social and cultural aspects of life, seems biased and collusive. Carlson's critique of the "engagement model" is, of course, in direct opposition to Berleant's ideas which I have said earlier that I clearly support. I would place my own interests somewhere between these conflicting positions.

Berleant and Carlson are obviously diametrically opposed in their positions on the appropriate model for aesthetic appreciation of nature and the environment. Where Berleant clearly states the need to collapse the subject-object dichotomy to integrate nature and culture once and for all. Carlson states that aesthetic appreciation is actually reliant upon the subject-object dichotomy, declaring that if you cannot define the object you can not achieve the goal of serious and appropriate aesthetic interpretation. I want to take a moment and consider an integrated subject/object experience and see if this is true. Five years ago, I was in Tokyo, Japan. I emerged from Shibuya railway station with my sense of personal space intact - and walked into a sea of humanity. I have walked and considered numerous cites around the world but nothing prepared me for the experience I was about to have. Waiting at the sidewalk for the lights to change, I stood in the densest crowd of people I have ever experienced. Piling up against the barrier of the street, the pedestrians blocked from crossing a road by rush hour traffic. As the light changed, I was amazed, amused and somewhat concerned when I realized that two opposing waves of humanity (literally thousands of people) were surging forward about to engage in the middle of a large urban crosswalk completely hemmed in by idling automobiles. As we moved forward, the crowd adjusted, ebbing and flowing like a school of fish and somehow making room for twice the population to occupy the same space. I stopped in the middle of the crosswalk and just watched in delight as this phenomenon engulfed me. Upon exiting from the station, I had entered into a public space where I, the appreciator, became part of a field of objects which I was experiencing. The subject/object relationship was completely dissolved. Yet I witnessed this event with a certain amount of disinterest, and was able to retain my sense of who I was and what it was outside of myself that defined the experience I was having. Indeed, not only did I emerge with my subjectivity intact, but I would submit that I was equipped to arrive at some serious and appropriate aesthetic interpretations exactly because of the collapse of the subject/object relationship. In comparison, an aesthetic philosopher with his subjectivity separate from the object of consideration - peering into this dynamic sea of humanity from a high rise building above this intersection, will likely miss important elements of the sensual, kinesthetic, social, cultural and scientifically informed experience of being on the ground as an object amongst like objects. Based on this experience, I can assume that the collapse of the subject-object dichotomy can occur at the level of experiential and conceptual understanding of the object without undermining the process of aesthetic appreciation. I would even suggest that a well trained philosopher (or artist in my case) can retain a sense of intellectual distance from the collective intent (commuting) of such an environment. These thoughts make me wonder if Carlson's defense of the subject/object dichotomy doesn't say more about the latent authority of critical appreciation as it relates to a separation between the making and thinking about artifacts than it does to the actual process of appreciation. With that said, I think its important to state that I agree with Carlson's position, but not his definition of the natural-environment model. In an increasingly complicated world where industrial residues from decades past have piled up to the point that they affect the global commons — the air, water and soils that sustain life — we must seek rigorous knowledge to inform the experience and appreciation of environment. Scientific knowledge is a primary choice to inform experience, but Carlson's decision to negate other forms of knowledge is short-sighted.

As a practicing eco-artist and theorist, I believe that we must allow for Carson's standard of significant and appropriate interpretation, carefully choosing the knowledge which informs aesthetics. But we must also allow for Berleant's aesthetics of engagement. Without a collapse of the subject-object relationship, we sit too far outside nature to understand the potential and moral imperative for integration.

Aesthetic-Systems and Health

Throughout this chapter, I have been clarifying the challenges that occur as we move from the industrial into the post-industrial and humanity, or culture, becomes aware of the pernicious impacts upon the essential commons that support life. In one century we have gone from the need to preserve and conserve to what I believe is an era where the ability to restore nature will become a paramount challenge. How do we appreciate (and act upon) the complex nature-culture systems of post-industrial nature? Traditional aesthetics would constrain us (the subjective viewer) to what can be known through direct visual experience of the object of contemplation, primarily the static formal qualities. Berleant's environmental aesthetic approach unifies nature and culture through the collapse of the subject-object relationship, and while Carlson's informs culture about nature through collaboration with science.

Another way to approach this question is to leave environment behind for a moment and go back to the question of aesthetics and beauty. Eaton in *"The Beauty That Requires Health,* (1997) suggests, "Aesthetic experience is marked by perception of and reflection upon intrinsic properties of objects and events that a community considers worthy of attention… anything that draws attention to intrinsic properties of objects and events can be described as aesthetically relevant." In this definition, she opens the door to senses beyond the visual and provides room for dynamic experiences by considering both objects and events. This definition is part of her ongoing work in philosophy and has been used in a number of her texts. I first came across it in a book edited by Nassauer called *"Placing Nature: Culture and Landscape Ecology."* Eaton's chapter " raises (but does not resolve) the integration of beauty and the perception of ecosystem health as a concept relevant to aesthetics. Admitting that the idea of health is general and poorly understood at the level of natural organisms and ecosystems, Eaton suggests a general policy to "…label ecological function with socially recognized signs of human intention for the landscape." She relates this idea to our learned ability to read the urban landscape for patterns that indicate abstract concepts like social or economic stability. She discusses aesthetic inventories and aesthetic examples as one way to inform the question of healthy natural systems. She concludes with the general idea that native flowers can bio-indicate soils without harsh chemicals and that slow and pleasing surface waters can indicate an intact and functioning natural hydrology of intact streams and porous surfaces.

Nassauer (1997) extends this idea in her own chapter, *"Cultural Sustainability: Aligning Aesthetics and Ecology."* She notes that ecological function is an increasingly dominant "intention" of public land but is still not part of the aesthetic that informs the design and management of private lands. Nassauer identifies the idea of "sustained attention" and the evolution of care (interface) as the path to new aesthetic knowledge and appreciation based on concepts of health. Her position is couched in rigorous knowledge of landscape ecology as a key concept in the aesthetic restoration of health in "settled landscapes". She provides a helpful comment in relationship to Carlson's over-investment in scientific knowledge. "Every possible future landscape is the embodiment of human values. Science can inform us; it cannot lead us."

What is environmental health and why should we care? There are two scales of health to consider - the organism and the ecosystem. There are three ways to think about health. One is the general perception of health, through knowledge gathered over time. We learn through regular interaction and experience to recognize a pattern of behavior that indicates the health, or illness of both organisms and systems. The second way to think about health is in terms of "a measure of the overall performance of a complex system that is built up from the behavior of its parts." (Costanza, 1992) The third way to think about health is in terms of autopoeisis, defined as a transliteration from two combined Greek words meaning self-making. The reason to care about environmental health is essential to Berleants concept of engagement and it is embedded in Carlson's idea of a natural environmental model. Understanding the lack of care and paths to change are the foundation of Naess' deep ecology, Plumwood's ecofemnism and Bookchin's social ecology. It is embedded in the struggle over the meaning of nature and its counterfeits which has roiled the philosophers and practitioners interested in restoration ecology. These environmental aesthetic theories emerge from a gnawing feeling that our natural and cultural systems are out of balance. That lack of balance is palatable and perceptible in experience but it lacks what Carlson calls serious and appropriate interpretation. I will discuss the general perception of health which, I believe we arrive at through pattern recognition and aesthetic analysis.

The relative health of a landscape, organism, ecosystem, even a technological construct, is a concept that most contemporary humans have experienced. While we may not be able to go into the details of systemic health, we share the zeitgeist of the term. We all know what a healthy person looks like. Many of us recognize factors that indicate a disrupted family unit. Failing communities, even failing management systems are obvious to most of us. Most of us even know when our computers or automobiles are getting "sick". We recognize health or the lack of health, through intimate multi-sensual experience and knowledge gained over time. Of course, there are numerous points of specific conflict in the application of the term health. Because of this, it requires a well-defined and carefully contextualized statement to provide a clear communication of the conceptual continuum in which health (or the lack of health) is being communicated.

The second aspect of health is in terms of measured performance. This following definition was developed as a result of a series of interdisciplinary meetings on ecosystem health at the Aspen Institute in Maryland [11],. "An ecological systems is healthy and free from distress syndrome if it is stable and sustainable - that is, if it is active and maintains its organization and autonomy over time and is resilient to stress." (Haskell et al 1992 Environmental economist, Costanza compares the knowledge of ecosystem health to human health. "Assessing health in a complex system -from organisms to ecosystems, to economic systems- requires a good measure of judgment, precaution , and humility, but also a good measure of systems analysis and modeling in order to put all the individual pieces together into a coherent picture." (Costanza 1992) Costanza proposes a general index of ecosystem health which measures the relationship between vigor, organization and resilience. Costanza points out that the range of knowledge (reference data) and diagnostic tools for human health far surpasses what we know about about natural systems. Without significant investment in research, it is still difficult to tell when we will be able to quantify a healthy natural environment. Returning to our aesthetic focus, I would argue that the intent of a quantitative system of measuring health in ecosystems, is primarily outside the realm of aesthetics. However quantitative health measurement could confirm or deny the value of pattern recognition as a relevant alternative approach to the question of health.

The third concept of environmental health is contained in the concept of autopoiesis, a relatively new idea only a little more than a decade old. L. Margulis and D. Sagan (1997) describe it as "to be alive, an entity must first be autopoietic – that is, it must actively maintain itself against the mischief of the world." This is a dynamic and reactive concept of health. The basic idea is that an autopoietic organism or an autopoietic ecosystem must have the ability to reproduce and sustain itself in terms of both structure and biochemical integrity. Autopoiesis can be perceived in terms of aesthetic pattern. It is easy to see when an organism has lost its integrity, harder but not impossible to see when an ecosystem has lost its physical, biochemical integrity, or when the organisms that define the system start to fragment and begin to lose their interactive complexity. Autopoiesis complicates both the general and the quantitative model of health, it embraces disturbance. This suggests a different sort of understanding and a new aesthetic model, which is not only dynamic, but also transactional.

Following Eaton, an aesthetic of health is, in my mind, an essential concept. According to the three models, health can be a general-aesthetic appreciation, it can be an expert-quantitative appreciation, and in the autopoietic lies the potential for the integration of the two. The first two are quite clear in terms of the "what" and "how" questions, posed by Carlson. An autopoietic aesthetic challenges the appreciator to embrace two entities (culture and nature for example) in relationship to one another. This adds a level of complexity in the decision to collapse or retain the subject-object relationship. Earlier I suggested that the "what" can be left to the appreciator; this works for the autopoietic as well as the first two models. "How" is a question that is less clear, defining what to consider would require a judgment about the state of the relationship as well as the state of the individual systems. This is a very specific and theoretical area of inquiry that is quite sophisticated. Understanding the science is a matter of attending to the patterns of relationship. There are a range of disciplines trained to perceive and clarify pattern. They should all have a voice in the development of this theory. Art and aesthetics can participate at the level of theory or at the level of interpretation. Establishing a voice is a matter of engaging oneself in a productive manner within the ongoing discourse.

Concepts and Tools to Aid Restoration Design

Can art also provide concepts, practices and tools for society and for restoring ecological spaces and consciousness? It surely can, although most of us in the United States have little understanding of art as a modern area of knowledge. The contributions of modern art are seldom discussed in our formal education programs, and are further undermined by a cynical response by the press and politicians to challenging contemporary work. At the same time, most of us with an interest in the environmental questions, have some sense of the historic contributions that artists and authors have made to the evolving idea of nature.

There is a good chance, that works by Muir and Thoreau; Lorraine and Monet; Claude, Church and Audubon have had some impact upon our understanding of nature. The contemporary artists Smithson (an original thinker in earth art) or the Harrisons (original thinkers in ecological art; see chapter 9 by Brown) are less likely to impact our educational or social realms of experience and learning. The meaning of art has changed radically in the last 100 years. At this moment in time, the word art standing alone refers to the artist's production framed by the authorizing reaction of institutional support, and an impact upon the viewer that demands intellectual and/or material consideration. What sounds like madness to the gatekeepers of another discipline is actually the strongest point of the fine-arts approach to knowledge. Without a deep institutional relationship to "knowing the world" in a specific way, we have the liberty to imagine, and dream the world in new ways. Theorists and philosophers as diverse as Kant (1886/1983), Gablik, (1984), Kosuth (1991) and Danto (1997) have raised issues about the end of art as we know it, but few outside the discipline recognize this condition. (our subject has been art itself) I would argue that art has reached the end of the reductionist pathway ahead of most other disciplines. I would argue that we are now reconsidering art and its relationship to knowledge, as well as its relationship to the world.

Conclusion: Art and Ecohumanism

In the introduction to this chapter I claim that we have entered a period of participatory ecology. We can no longer take for granted natures ability to maintain itself in the background, while humanity lives in the cultural foreground. We could describe the innocence of the industrial period, where nature was assumed resilient, and constantly functioning in the background as a second Eden. In the original garden, Adam and Eve only had to bite the apple, to learn what they had lost, today within what was what I would call a second Eden, we have begun to understand that we have consumed the entire tree, poisoned the soil and changed the climate of the garden. Consumption is once again, the path taking us to the point, where we realize what have lost. Moving away from Eden-1, we found cultural awareness, which we believed was separate from nature. Moving away from the conditions that were Eden-2, we may find a new awareness, a fusion of humanity with nature, that results in a single consciousness. We could call this paradigmatic shift in human conscious ecohumanism [11] (Tapp, 2002), whereby we become responsible for the restoration, healing and long term health of nature as an extension of the human condition. I would describe this as the emergent condition of Eden-3.

The question of nature is increasingly addressed by a range of radical interests, radical in their intent to change the social relationship to nature. The theorists and practitioners that have informed this section describe specific problems of the nature-culture relationship. Jordan describes restoration as an intimate relationship, where we become privy to "secrets" about nature. Elliot and Katz suggest that there is nothing that can be done with those secrets. Light and Foster provide us with critical insight on integration, and how it differs when viewed from either end of the statement, as a culture of nature and a nature of culture. Bookchin, Naess, Merchant and Plumwood provide us with radical frameworks to reconsider our beliefs. Berleant and Carlson provide us with philosophical frameworks that either transcend the separation of nature and humanity, or clarifiy the import of outside perspective (objectivity) when we advocate for nature. Costanza, Eaton, Margulis and Sagan provide us with an overview of the tools that are evolving as we grapple with the nascent concept of the health of nature.

The ecological artist has incredible potential to participate in the post-industrial project of ecohumanism. (see Chapter 5 by Light.) Humanity has lost its relationship to nature. Nature has faded into the background during the industrial period. What is the role the arts can play in response to this loss? Artists with a knowledge and passion for new cultural concepts, have always been on the forefront of metaphorical and symbolic knowledge (see Chapter 8 by Brown) I would argue that metaphoric knowledge is the root of a paradigmatic understanding. Contemporary artists are comfortable with complex ideas and their affect on human perception. Artists understand the impact of systems with a good interface, as well as what happens when interface is lost. Perception can be enabled or constrained by interface and human values follow perception, framed within known concept models. Like the contemporary aesthetic philosophers, artists have to slip some of the bonds of history, and think carefully about how to define interdisciplinary practice and what it means to act upon these ideas within culture. In a culture dominated by science, which expands and defends what we know based upon a foundation of knowledge, the arts have to develop new critical and strategic tools to act upon society. We need to create a supportive interdisciplinary community of creative individuals that are committed to, and take responsibility for, positive shifts in the "culture of nature." We also have to be responsible for the knowledge and impact of our work across disciplines. In the interdisciplinary model, we find new reasons to think about the efficacy and impact of the artist. At the same time we must consider how we make these arguments in the context of a discipline that denies a foundation approach to knowledge. The cultural value of art lies in its ability to question the canon, rules, principles and standards that confine the thinking of other disciplines. The artists unorthodox approach to knowledge often opens unexpected doors. Shedding daylight upon options, issues and solutions that would not be considered or pursued through more tradition-bound disciplines, conventional social programs or political and economic institutions. This is not an issue of comparative value, but rather one of complementary value.

Chapter 10

Constructing Restoration Ecologies: Nature, Aesthetics, Sites and Systems

Louise Mozingo

Like ecological art discussed in the last chapter, landscape architecture conjoins the expression of aesthetics and the design of nature. During the last decade restoration has emerged as a promisingly vital part of the landscape architecture profession. It has engendered energetic commitment among practitioners and a satisfying level of public approbation. The design of restoration landscapes, from the modest replacement of a patch of lawn with native grasses to the ambition of the Everglades project, the largest wetland restoration in the world, distinguishes the discipline of landscape architecture as it enters the 21st century. The context for this is, of course, our collapsing global environment and the proposed redirection of human priorities towards sustainability. Restoration landscapes constitute an essential part of sustainability—the reformation of landscapes debased by human action.

This renewed professional fervor is understandable not only as a conscientious response to global circumstance but also as a serendipitous evolution of landscape architecture at a particular moment in history. While landscape architects come armed with talents especially suited to the accomplishment of restoration landscapes, the design of restoration landscapes engages landscape architects in a way that brings assurance and purpose to their endeavors. (See survey results examined by Ryan in following chapter.)

After the disorientations of the modernist design era, landscape architects can once again connect their work to the heroic legacy of nineteenth century American landscape architecture. The modernist onslaught dismissed the Olmstedian landscape as hopelessly outdated obscuring the superb restoration work initiated by Olmsted, and carried forward by Eliot, Cleveland, Simonds, Jensen, and the younger Olmsteds. Without resorting to historicism, resonant historical models inform new projects, placing this work in the landscape altogether more convincingly than the de-contextualized theories of modernist design.

Restoration landscape architecture reinvigorates the ideals of the American public landscape. The post war dispersal of urban density coupled with the expanded availability of private outdoor space diminished the use of collective spaces; uncertain purpose and indifferent design for the public landscape followed. (Jacobs 1961) Restoration has catalyzed the rehabilitation of many public landscapes and the creation of new ones at a scale and ambition not seen in several decades. The inclusion of restoration landscapes in the public realm has sparked design creativity that for decades seemed confined to the non-public realm.

At their best, restoration landscapes use science comprehensible at the scale of human perception. While McHargian science, in practice, communicated more about what not to do at an abstract scale, the science of restoration actively and obviously shapes landscape space. The physical science in restoration landscape design adds both concrete direction in wielding landscape architectural expertise and the currency of measurable validity—either the ecological processes are present or they are not; this validity markedly contrasts with the elusive metrics of social science that underlay the practice of landscape architecture in much of the twentieth century.

The educative aspect of restoration landscapes focuses the social goals of landscape architecture toward accomplishments that the landscape can actually produce. While the broad remaking of society expected by modernists did not come to pass, nor did the empowerment promised by the more radical advocates of the community design movement, the landscape can indeed be didactic about the landscape. Societal comprehension of the intertwined aspects of human action and landscape health is possible in the experience of restoration landscapes. It forms a congruency between means and ends that was absent in the modernist landscape and the social goals it proposed to address.

This hopeful position for restoration, as a significant contribution by landscape architecture to sustainable human practices and, consequently, a welcome seriousness to the profession itself, also requires vigilance as to the presumptions behind the design of restoration landscapes. In the initial enthusiasm for the possibilities of restoration and in the fray of practice, these assumptions have been present but not necessarily explicit. Yet they emerge as contentions that assail the practice of restoration: the meaning and definition of nature; the significance of landscape aesthetics; and the relationship between landscape site and ecological system.

The Meaning and Definition of Nature

The recent philosophical and ethical contentions over the meaning of nature as it relates to the theory and practice of restoration recalls a passage from the eighteenth century Chinese classic *The Story of the Stone* (also known as *The Dream of the Red Chamber*). In it too the meaning and form of nature is the subject of vociferous declaration. The novel's hero, the precocious son of the well-to-do family, accompanies his father and a coterie of literary gentlemen on a tour of the family's new garden. At one point the son questions the father's preening admiration of the garden's "natural beauty." The father proceeds to admonish his son: "What you know of the beauty that lies in quietness and natural simplicity? This is a consequence of your refusal to study properly;" and the son responds, "Your rebuke is of course justified father but then I have never really understood what the ancients meant by 'natural.'" The father further rebukes his son: 'Why fancy not knowing what natural means—you who have such a good understanding of so much else. Natural is that which is of nature that is to say, that which is produced by nature as opposed to that which is produced by human artifice." The son seizes upon his father's words "There you are you see!...when the ancients used the term natural I have my doubts about what they really meant. For example, when they speak of a natural painting I can't help wondering if they are not referring to precisely that forcible interference with the landscape to which I object: putting hills where they are not meant to be, and that sort of thing. However great the skill with which this is done, the results are never quite..."The father interrupts, "Take that boy out of here!" and the boy flees (Xueqin 1973).

This passage from both another culture and age gives us perspective that the current scholarly controversies over the "natureness" of restoration are imbedded in perpetuating discussion not resolved definition. Like the son in the Chinese classic, critics contend that the "nature" of restoration is a fundamental misnomer—nature can only be the result of ecological process extraneous from human intervention—and, significantly, that this misinterpretation is dangerous because it facilitates ongoing destruction of true nature in the soothing illusion that nature can remade (Elliot 2000 [1982], Katz 2000 [1992], Katz 2000). Defenders contend that nature has never been devoid of human intervention since the advent of the species—the problem is instead the scale, manner, and results of intervention—and the restorationist agenda is a pragmatic, ethical, and "healing" response to current environmental conditions (Attfield 2000 [1994], Light 2000, Jordan 2000[1994], Foster 2000). From the point of view of an on the ground restorationist these oppositional contentions, bizarrely enough, each have resonance. Restoration landscapes may indeed, perhaps even usually, fall short of the diversity and complexity they are meant to restore yet I would contend that the awareness of this very difficulty has increased the impetus towards the preservation of functioning landscape systems.

Both of the above positions skirt how definitions of nature come into play in restoration projects as I have experienced them. To the general public nature is, rather simply, predominantly green landscapes (Kaplan and Kaplan 1989, Kaplan, Kaplan, and Ryan 1998; Gobster and Barro, 2000). Depending on the community, this definition may be refined (but usually is not) to include the presence of regionally indigenous species. To restoration scientists working in North America, nature is an estimation of the given landscape before European colonization—nature defined by a moment in history. Upon scrutiny, it is a definition enmeshed in liberal politics and the condemnation of the Western imperialism that destroyed the imagined prelapsarian idyll of Native America.

All of these definitions of nature are instrumental to their contenders and at some level valid. From purist to practical, they also share a separatist, if not oppositional, view of nature and culture—their contentions lie in the optimal portions of nature and culture. These working definitions of nature illustrate what many scholars have already discussed—that the idea of nature is a cultural construct mutable over time and between societal groups (Glacken 1967, Merchant 1989, Cronon 1995). Restoration conjoins this variable yet potent cultural construct to the presumptive certainty of science—an ungainly attachment at best, resulting in almost inevitable confusion.

The diversion of the discussion of restoration into a discussion of the elusive definition of nature misses this most salient point: the definition of nature is shaped by human action—specifically by human action in the landscape. Today we are assembling and building a new set of landscapes that are shaping the contemporary meaning of nature. We are insinuating restoration landscapes, constructed ecologies, into our everyday environments, into the "daily paths" that form and iterate cultural norms.[1] An exposed creek appears in the middle of a neighborhood park and captures errant tennis balls; the gleeful dash from parked car to shoreline requires an extra trudge up and over the carefully delineated path through restored dune grasses; the newly mucky edge of an estuary of what had been fill land near a busy airport changes colors with the seasons, stinks, and shows off the startling elegance of egrets. The restoration precept that brought about these landscapes was an idea; the stormwater treatment wildlife pond at the center of a subdivision is a quotidian reality.

The process of shifting and shaping the definition of nature implied by these restoration landscapes was not initiated in the last two decades when the restoration of conservation biologists has come to the fore—rather American landscape architecture of the public realm has been doing this since Central Park. Most folks perceive Central Park as nature, perhaps not wilderness, but certainly nature. This is true even of Golden Gate Park in San Francisco, a landscape of even more contrivance than Central Park—a green bucolic swath concocted upon the sand dunes of the given landscape. The artifice of these built landscapes could evoke a skeptical if not condemning judgment against its set piece illusion, its theater. But just as theater, at its best, allows us to shift the accepted parameters of our existence, to consider new ways of understanding the world, the contrivance of American public landscape architecture led directly to the advent of restoration landscapes. The popular acceptance of a contrived version of nature led to the possibility and the implementation of the fundamental models of restoration in the United States—Olmsted's Emerald Necklace and the reforestation of Biltmore; and Charles Eliot's conception of the Boston Parks and Reservation system which included dune, river, forest and meadow restoration and preservation. These east coast exemplars played out in the Midwest—Horace Cleveland's park system in Minneapolis; the Chicago reservations and the prairie landscapes of O.C. Simonds, Jens Jensen, and Willem Miller. The midwestern legacy of American landscape architecture of the public realm is directly linked to the Chicago restoration efforts that sparked such heated controversy in the 1990s (Helford 2000).

Landscape architects have been building artifactual nature as part of the public realm for at least a century and a half and, in the process, shifting and shaping the definition of nature so that restoration landscapes were conceptually possible. In short, the artifactual quality of American landscape architecture has allowed us to imagine the possibility of restoration.

That this artifactual nature, first adopted in the public realm towards the social goal of recreation, has been remade for ecological goal of restoration is not dubious. It is at once both a pragmatic and idealistic response necessitated by the circumstances of restoration. First, where and how these restoration landscapes occur forces their artifactual quality. Restorations take place in landscape fragments of larger systems that have been spatially deformed in the process of degradation; they are proximate to and even enmeshed with closely guarded human settlements. This is where restoration works, with these compromised remnants that require significant manipulation to restore ecological vitality. Second, these restorations involve negotiated goals based in a complexity of agencies and communities—often vociferous in their advocacy—that contributes to the necessity of artifice. Third, the artifice of restorations can have measurable ecological benefits by improving water quality, habitat, and biodiversity—they are constructed, but ecologies nonetheless.

And lastly, restoration landscapes are valuable precisely because they do not rely on an oppositional position between culture and nature in which, as Brunson (2000) states, "the ultimate loser in all such conflicts is nature." They rest on both deliberate human intervention and spontaneous ecological process. Rather than conceiving this as obscurantism or disguise, they restore the conception of nature that held sway from ancient Greece to the eighteenth century—that is, humans and their culture are a part of nature. For most of Western history, the cultured landscape prevailed as a version of nature, a "second nature" to use Cicero's term, an idea discarded only with the advent of industrialism (Glacken)[2]. Indeed, some scholars contend that a separatist view of nature facilitated and enabled the environmental destruction of industrialism (eg. Glacken, Merchant, Brunson). Restoration thus also connects to the hopeful precepts of urban ecology, a contemporary and constructive re-emergence of the Ciceronian idea.

The Significance of Aesthetics

Rather than resist the artifactual reality of constructed ecologies, this reality should be utilized to convey a new landscape order; the theater of these landscapes should be emphasized rather than diminished so that this scatteration of projects, the quotidian restoration landscapes, coalesce as an environmental vision. The success of constructed ecologies as an alternative environmental vision depends upon regarding them properly as landscape designs. They require the powerful results of iconic designs of the caliber, for example, of Versailles. If constructed ecologies are to establish commanding ground, both figuratively and literally, it must be as unapologetic as Versailles; to have the sense as Louis XIV did that the stroke in the earth was the most apparent and immediate manifestation of a reorganization of environmental power. Iconic landscape designs, characterized by notable aesthetic quality, create attention; they resonate over decades, even centuries. They are admired, preserved, and, most tellingly, imitated. They manifest and promote environmental change. They become integrated into the hierarchy of places a culture values most.

By becoming iconic, constructed ecologies can most effectively redefine good landscape form in response to the pressing concerns of environmental degradation. It addresses both the artifactuality and the communication that these landscapes should convey about the "public ecology" of restoration (Hull and Robertson 2000). The positive aesthetic experience of "in the ground" built projects that encompass new ideas of good landscape form can positively promote change. However, examining constructed ecologies in tandem with landscapes of notable aesthetic quality—landscapes within the tradition and history of landscape architecture—elucidates the challenges of synthesizing the persuasive aesthetic precepts of iconic design and constructed ecologies. Conceiving of constructed ecologies as landscape design requires an explicit understanding of visibility, temporality, reiterated forms, expression and metaphor. Like the aesthetics of ecological art discussed in the previous chapter by Collins, this understanding suggests the means of a culturally integrated aesthetic of the restoration landscape.[3]

Visibility

To promote ecological design, making it a perceivably visible landscape experience is essential because the visual experience creates comprehension of a shift in environmental priorities. Yet, the perceptual subtlety of ecological landscapes can make them difficult to perceive—and to care about. Very often constructed ecologies incorporate "visibility" (Hough 1994) or "observability" (Thayer 1989, 108) of ecological phenomenon as simple revelation, with little attention to context or point of view—it is there so it must be visible (Nassauer 1992, 244). The West Davis Pond, a water treatment wildlife pond in Davis California, could easily be mistaken for leftover lowland instead of the considerable piece of artifice that it is.

Part of this perceptual subtlety has to do with the way we have built our environments in the industrial era—with little attention to viewpoints and pathways that focus our attention as we move through the landscape. San Leandro Creek, is a very typical urban stream wending its way through the city grid of East San Francisco Bay. It is, for the most part, illegible, not because it is inherently ungraspable but because the entire urban fabric is set to ignore and obscure it. In comparison Strawberry Creek as it runs through the Berkeley campus provides select pathways and viewpoints to illuminate the systemic quality of the creek. (an Olmsted design at its core by the way) The continuous thick line of trees is obvious; approach its banks, and the sudden sweep downward in elevation and watery shadow palpably marks its presence; move within the creek, and the vegetative enclosure and the ready perception of upstream and downstream flow connects it to the unavoidably apparent East Bay Hills urban forest and San Francisco Bay waters. Vital ecological systems can be understood as ground level experience given attention to pathway and viewpoint.

In the service of visibility, constructed ecologies can also exploit the power of contrast between cultural and ecological systems. Boston's Emerald Necklace, and in particular the Riverway and the Fens designed by Olmsted during the nineteenth century, is just such an impelling juxtaposition. Imperiously shaping an expanding city (Boston grew around it for the next several decades) it is aesthetically arresting because of its surrounding city context not in spite of it. Bringing together ecological and cultural systems engages rather than repudiates the cultural milieu where most restoration design takes places. The Emerald Necklace vividly demonstrates that structuring the cultural landscape is as important as restoring and preserving the ecological landscape, with the implication that their conservation means are different, yet ends are the same.

Temporality

A fundamental issue in the design of constructed ecologies concerns the consequential acceptance of ecologically necessary landscape change. There will be times that the meadow is dead, the wetland birds flown, the urban forest burnt, the surface runoff channel filled with stagnating water, and the flooded stream will blast an eloquent arrangement of rock and carry a favorite sycamore downstream. The public certainly, but many landscape designers as well, consider landscape change not as a vital, imaginative force but as frightening or disappointing one. Landscape changes generated profound distress among Chicago area citizens who lived near prairie restoration efforts undertaken in the early 1990s. The ensuing protest over these changes created resistance further restoration actions and stopped the project (Gobster 2000).

This indicates, that while moving beyond the fixed vision imposed upon the land is essential, some fixed points are integral to an intelligent conception of ecological design. The desire for constancy and the necessity of dynamism evolve into the more subtle sense of continuity. Some parts of the landscape will be the same year after year and are places appropriate for overt aesthetic statement and that these introduce and present the fluctuating part of the landscape—in the way that a boardwalk over a marsh can show us a landscape that is on occasion creepily muddy and on occasion serenely refulgent.

Reiterated Forms

Another challenge in understanding restoration landscapes as iconic design is that the charge of ecological design is to respond to the particularities of site processes. Yet throughout history, a vocabulary of forms, reiterated across time, space and cultures, characterize iconic design aesthetics. Aesthetic patterning may be ecologically arbitrary but it is culturally deliberate (Rapoport 1990).

The opportunities for reiterated form are actually ample in ecological design. Most of these interface restoration landscapes of degraded ecological systems require extensive landscape restructuring that can be the locus of design attention. In the urban creeks of California—grazing, farming, dam building, and tree clearance—have resulted in dramatic downcutting of streams. The biodiversity fostered by a shallow meandering creek is long since gone. In these creeks, biodiversity can be achieved only through the introduction of constructed intrusions in the creek channel. These should reflect a level of reiterated patterning that reinforces the perception of the creek restoration as a systemic extension across the landscape and inspires the collective regard that these landscapes deserve and require. Part of this is craft—the details matter. What does it say when "Flows to the Bay" sign is spray painted—this is the medium of graffiti. In no culture are places of collective esteem coarsely made, rather they are made to the highest level of craft that a society can muster.

In the implementation of constructed ecologies, we should not forget that many of the familiar and beloved design forms of the Western landscape tradition began as eminently ecological elements, appropriate to their time and place. The parterre and the runnel as irrigation forms for the parsimonious use of water, the grove and clearing as managed wildlife habitat, the cascading renaissance pool as a sparing system of water reuse, the allee as cover from rain and sun along a country road reducing runoff and reflected heat, the city block as thrifty use of urban land. However much regarded as imposed upon the land or as purely visual elements, they made some ecological sense in their day, and to rediscover their ecological value and function must be part of the aesthetics of constructed ecologies.

Expression

In considering expression in the ecological design, the imaginative strength of landscape artists provides clues to both the necessity and means of expression in ecological design. While the land art pieces themselves are certainly provoking, the act of their presence on notably anonymous and remote landscapes, lends these landscapes a vividness hitherto unseen (Baker 1983). This has direct application to restoration sites, which, like the West Davis Pond, are at times surprisingly anonymous, coy about their significance in a broader ecological system, and reticent about their inherent vitality. Ecological design, as part of a public dialogue in reordering environmental building, needs the expressive marking of art—it is perceptually facilitating. It lets us know that we should attend to this place (see chapter 8 by Brown and chapter 9 by Collins for a related discussion of the role of artists in restoration).

Metaphor

Ecological design cannot be a merely empirical product which that does not consider elevated, or beyond the self perception, denying the opportunity for cultural connection and reflection (Nassauer 1992, 1997). The manifestation of metaphors in constructed ecologies engenders landscapes that are not just ecologically valuable but socially valuable, recognizing that the two are inextricably intertwined. By engaging metaphors in the construction of ecological function restoration landscapes move beyond nature gazing and become places that are necessary, not optional, parts of human life.

At the Woodland Cemetery in Stockholm, Sweden, G. Asplund created a place that is not only environmentally valuable—a restored forest on a degraded quarry site—but also a place of meaningful human ritual—a place to bury and visit the dead. This is a place of mourning and regeneration, both spiritually and ecologically symbolic of hope and continuity. The web of process weaves together the ecological and the cultural. Human interaction with this ecological landscape moves from the optional to the necessary, safeguarding both ecological and human values.

A central question in discussing the potential for metaphor in the ecological landscapes is whether there is a basis of "ecological literacy" which can inform and shape a new set of metaphors (Orr 1992). The next generation will be more ecologically literate but they will be literate about ecology in a very unprecedented way and not as Aldo Leopold might have expected. Ecological literacy is being generated from everyday urban context, rather than rural or wilderness context typical of the environmental education of previous generations. Environmental learning now takes place in urban creeks and city open spaces, in marshes along freeways and the urban forests along our streets, and literally in the gutter. This culturally and socially inclusive form of environmental learning vastly broadens the possibilities of "shared meanings" that is at the heart of building sources of metaphors. This new ecological literacy, borne by both education and, more importantly, by experience within an everyday landscape, can build new language for landscape metaphor (see chapter 11 by Ryan). The restored ecological systems within the city cannot expect to have the ecological caliber of those within the preservation landscape but they have much more power to change culture in their immediacy and proximity. The consequences of restoring and making manifest ecological systems in the city extend beyond measurable ecological indices.

Visibility, temporality, reiterated form, expression and metaphor are all ways to convey the meaning and significance of constructed ecologies. Part of that meaning is the artifactuality of restoration landscapes and its propagandistic potential to manifest the radical reorganization of environmental priorities that restoration design implies, to build a "second nature" of restoration.

The Relationship Between Landscape Site and Ecological System

By bringing to bear the seductive sensuality of landscape design to restoration landscapes there are significant risks. Most fundamentally, we must expect that the fragments of ecological systems that we are struggling to deal with will be the first interventions in what will become a systemic coherence. At a 1995 conference on landscape ecology Richard Forman, author of the fundamental text *Landscape Ecology,* assessed the ultimate effect of a large constructed wetland as "tinkering."[4] Most members of the audience (myself included) audibly bristled at the characterization of a hopeful and elegant project, yet the point was arguable. In terms of fundamental environmental improvement, the constructed ecologies of fragments alone may be a rearrangement of deck chairs on a sinking ship. If landscape architecture's contribution is to design those deck chair arrangements really nicely then we are, perhaps inadvertently, perpetuating the problems we are earnestly trying to address. Much fragment restoration is now justified by its "educative" potential; the time has come to move beyond the educative, to building measurably environmentally constructive landscapes.[5]

The successes of restoration per se may elide the more complex, much more difficult environmental choices that are in front of us. Put simply, I worry about the restoration volunteer that lovingly plants riparian forbs but drives by himself in the Ford Explorer to the work party—that the immediate satisfaction of planting and making eases our conscience into thinking that we have done enough, as individuals and as nations. A systemic restoration of an urban stream with native vegetation can be an appealing and sensual version of nature but will be of very limited positive effect without restructuring other parts of the landscape. The immensely compelling aesthetics of European post-industrial restoration landscapes beg the question of where those very dirty industries now are. Some perhaps are obsolete, but most have reappeared in "emerging" economies with consequent environmental destruction. This is a global version of masking the hovels of destitute tenant farmers with an artful grove, hiding the slave tended rice fields behind the allee of live oaks, or the PCB riddled transformer behind the neat hedge: the post-industrial West as a continental Reptonian landscape in all its solipsism; a "grey world, green heart," to use Rob Thayer's insightful description, at planetary scale.(Thayer 1994) Restoration landscapes need to more explicitly connect to the other necessities of ecologically constructive design—density, recycling, energy conservation, closed loop production systems, *and* preservation of ecologically functioning landscapes, near and far.

For landscape architecture the worst risk of all is the co-opting of restoration landscapes as merely visual style without attendant ecological function. The marketing of Crissy Field chocolate tin presumably raises funds to support restoration efforts yet it is reductively emblematic, made all the more so by the notable failure of the constructed marsh as a functioning ecological system.[6] Some of this might seem slightly silly and harmless, but multi-national corporations, agents of massive environmental degradation, also deploy the trope of restoration. The 1992 Boeing Longacres Campus outside Seattle incorporated a large constructed wetland at the center of its corporate headquarters. The public story that the Boeing Longacres Park tells is of environmental responsibility, of abiding by, even extending beyond, environmental regulations for the public good, of literally having environmental concerns at the core of its corporate life. Yet in the same year that construction began at Longacres *Fortune* singled Boeing out as one of ten egregious corporate "Environmental Laggards" noting beyond the norm increases toxic, chemical, and solid waste production (Rice 1993, 122).

The transformative power of constructed ecologies lies in their unified vision of human and ecological process, deployment of persuasive, even propagandistic, aesthetic principles, and truly constructive ecological outcomes. Through my wincing caution, I remain convinced of the promise of building constructed ecologies, of broadly reshaping the way we build as ecological design. Several years ago my students and I completed a master plan for a seventy-acre track of hillside open space newly acquired by the city of Oakland.

For many years the site received the regular attention of gang members who amused themselves by setting fires to the slopes in the dry season. As an initial step in the process, a Berkeley colleague from the College of Natural Resources, who had long studied the fire cycle of our region through a distinguished career, accompanied us on the field trip to the site. As he and I surveyed the site I commented that we were going to have to "do something about the fires;" his reaction was "people will have to get used to them." I thought he was plumb crazy—the fires were too scary, the charred hillsides too ugly, the community would never accept it, the homeowners next to the open space would freak out, the city would never allow it. Later, one of my students discovered that because of the periodic fires, the site contained the largest extent of bunch grasses in the entire East Bay and one of the largest in the entire Bay Area—a treasure trove of native species. Once the community knew that, we all discovered that a plan could be worked out to let the fires continue as part of the master plan. With good science and good design we all stepped in a different, and salutary, direction of human action in the landscape. The gang members left an enviable, if inadvertant, legacy.

Chapter 11

Understanding the Role of Environmental Designers in Environmental Restoration and Remediation

Robert Ryan

Introduction

The proliferation of degraded land and water bodies that has resulted from the modern industrial era is staggering. In the United States alone, the U.S. Environmental Protection Agency estimates that approximately one-third of nation's rivers and other water bodies which do not meet federal standards for fishing and swimming (U.S. E.P.A. 1997). Furthermore, "it is estimated that 75% of the 175,000 known waste disposal sites in the country may be producing hazardous waste chemicals that are migrating into groundwater sources" (Botkin and Keller 1995). There is a tremendous need to restore and rehabilitate these "brown fields and gray waters" for future human use, as well as repairing the serious rents in the ecological fabric of the planet. Landscape architects along with other environmental designers have played an important part in cleaning up brownfields and other toxic sites, finding new solutions to treating urban stormwater and other non-point source pollution, and worked toward repairing and restoring wetlands, streams and other water bodies (see preceeding chapter by L. Mozingo). However, the bulk of research and publications about environmental restoration and remediation has come from the scientific and engineering fields (e.g., Ecological Restoration Journal, Jordan et al. 1987). Furthermore, those researchers that have looked at the social implications of ecological restoration focus their efforts on ecosystem restoration projects in which restoring nature to some pre-settlement condition is the main priority (see Gobster and Hull 2000), rather than dealing with the harsh realities of severely polluted sites and waterbodies that must also serve future human uses.

There is a need to understand more about the role that landscape architects and other environmental designers play in environmental remediation and restoration projects. As has been suggested by several researchers, environmental restoration not only repairs the environment, it also reconnects people to the land (Ryan et al. 2001, Jordan 1989, R. France in Intro.). According to landscape architects and other environmental designers, what are the most important goals for ecological restoration and remediation? What is the role of landscape architects in these multi-disciplinary projects? What are the skills and knowledge base that must be brought to bear to be effective in environmental restoration and remediation? And, just as designers shape these projects, how will continued work in this field of practice shape the future of the landscape architecture profession? This chapter will present the results of a survey of landscape architects and other environmental professionals about environmental restoration and remediation, and presents new insights for both environmental restoration and design education and practice.

Environmental Designers' Viewpoints about Ecological Restoration

A review of the literature reveals several themes about the goals of ecological restoration and remediation projects. According to environmental philosopher Holmes Rolston III (1994), the goal of ecological restoration is to facilitate the healing process. By tapping into nature's power to heal itself, ecological restoration can reinvigorate natural processes to sustain a healthy, functioning ecosystem. Environmental artist M. Ukeles describes her proposal for the Fresh Kills landfill on New York's Staten Island: " 'This place could be a site of transformation, where people could see our power to take something that was so degraded, and such a hard thing to bear, and heal it.' But to heal is not to erase, according to Ukeles" (O'Connell, 2001). Landscape architecture and engineers should not hide previous environmental degradation, but create sites for transformation — places that reveal humans' power to heal a place. Furthermore, these designs should be informed by the flow of life: people, water, and even garbage.

Brown and others (1998) suggest that ecological design should improve the ecological health of a site while revealing the dynamic quality of nature, such as making the movement of water more legible. Interpreting ecological processes and relationships is seen as a key aspect of ecological design, which includes environmental restoration projects. In the ground-breaking exhibit of ecological design projects, entitled, "Eco-Revelatory Design: Nature Constructed/ Nature Revealed" the conference organizers describe several themes that unify this body of ecological design work. They see abstraction of nature and natural processes as much a part of ecological design as it is a part of garden design. This includes simulating nature and natural processes in order to create a deeper caring for nature, as well as plan for new uses. In addition to exposing natural processes, these landscape architects, like Ukeles, suggest that exposing infrastructure processes (i.e., stormwater, trash removal) is another important aspect of ecological design projects. Another aspect is reclaiming and remembering the natural and cultural heritage of a site, which includes preserving and restoring significant landscape features. Finally, ecological design should change the public's perspectives on how they interact with a landscape or even the environment as a whole. This last theme, which has as its goal building environmental ethics through environmental education, appears to underpin many environmental restoration efforts involving landscape architects. While not discussed widely in the environmental design literature, scientists and other practitioners of environmental restoration have suggested that environmental restoration provides a fertile ground for conducting research on natural systems (Jordan et al. 1987).

Landscape architect J. Nassauer (1995) proposes that landscapes, which promote ecological health, should appear intentionally designed and managed. In order to promote landscape stewardship, designers must create landscapes that are ecologically-beneficial and also aesthetically pleasing to the general public. Creating orderly edges and boundaries to otherwise "natural-appearing" restoration projects helps to create the "cues to care" that Nassauer states are vital for public acceptance and stewardship of these places. My own research on preference and attachment for urban natural areas in Ann Arbor, Michigan, which were undergoing ecological restoration efforts, suggest that people living near urban parks in particular are much more inclined to desire these "cues to care" than general park users. Furthermore, the public's perceptions may be very different than those held by landscape architects and other park planners who think urban natural areas should be restored to more natural-appearing, pre-settlement condition (Ryan 2000). It would be interesting to determine if these professionals' attitudes toward restoration hold true in more degraded landscapes, such as brownfields and landfill areas where existing site conditions are so much more dramatically altered.

In a study that looked at the public's attitudes toward ecological restoration in Chicago's forest preserves, Vining and others (2000) found that those who supported ecological restoration work gave reasons such as improving species diversity, reclaiming the area's natural heritage, environmental education, aesthetic beauty of restored ecosystems and removal of invasive, non-native species.

Environmental philosophers have argued the validity and definition of the term ecological restoration. W. Throop (1997, 2000) argues that narrow definitions of ecological restoration that focus on restoring a site to pre-human settlement are particularly problematic. For example, should the goal be to restore a site to a pre-European settlement condition or prior to Native American settlement? There is often a lack of information about earlier ecosystems and in many cases, the surrounding landscapes has altered so significantly that a particular site is influenced by radically different environmental influences. Furthermore, a narrow view of ecological restoration presupposes that a "wilder" restored ecosystem is of more value than another type of "new" ecosystem. Other authors in this book, such as A. Light (Chapter 5), take this discussion head-on. It would be interesting to see where landscape architects and other environmental designers fall on this spectrum of ecological restoration, from simply repairing some previous environmental damage to attempting to replicate some more "pristine" pre-settlement ecosystem.

Sociologist D. Lodwick (1994) argues that environmental restoration must have at its central goal making cleaner, safer environments for all people, especially inner-city minority groups who are disproportionately surrounded by degraded land and waterbodies. If the environmental restoration movement does not pursue more human-centered goals, it will not be embraced by the general public and will remain the pursuit of upper and upper-middle class environmentalists for some idealized form of nature, the untouched wilderness.

This review of the literature revealed multiple reasons for engaging in ecological restoration including healing degraded landscapes, abstracting nature and natural processes, aesthetic expression, promoting environmental stewardship by reconnecting people to the land, environmental education and research. The goal of the present research study is to determine if these perspectives on environmental restoration and remediation were shared by the practitioners, academics, and students who are engaged in this type of work.

Methods

Study participants

The sample for the survey were participants at the "Brown Fields and Gray Waters" Conference sponsored by the Department of Landscape Architecture at the Harvard Design School. This international conference was designed to "critically examine the design options and planning procedures for remediating and restoring degraded landscapes." Surveys were distributed to all 253 attendees, 94 completed surveys were returned for a response rate of 37%. The participants reflected the interdisciplinary nature of the conference and environmental restoration field. Approximately half of the sample were landscape architects, whereas the remaining participants coming from the fields of architecture, planning, engineering, environmental sciences, public policy, government, art, and literature. The sample also covered a range of types of practice with half of the sample from the private sector and the remainder divided between public sector, academia, and students. In general, this was an experienced group of professionals, who had worked an average of eleven years in their respective professions. The sample came from a wide geographic area with only a third of the sample being residents of Massachusetts, half were from other states and the remainder were international residents. The study sample was almost evenly divided by gender, male (56%) and female (44%), and relatively young with over half of the sample in their 20's and 30's, and the remainder in their 40's, 50's, and 60's.

Survey instrument

The survey instrument was designed to understand how landscape architects and other professionals perceive the issue of rehabilitating land and water bodies that have been severely degraded by human development. Participants were asked to rate the most important goals for ecological restoration projects from a list of 38 items derived from the literature. These goals included topics such as cleaning the environment, social benefits, human health and enjoyment, urban renewal, artistic expression, natural and cultural interpretation, and environmental education. The survey asked participants about the role that landscape architects should play in rehabilitating degraded landscapes and waterbodies, the skills and knowledge that landscape architects need to have to be effective in this type of practice, and the degree to which these key skills are currently being taught in existing landscape architectural education programs. Participants were also asked about their own knowledge and experience with various aspects of ecological restoration and rehabilitation. For the majority of questions, respondents were asked to use a 5-point Likert scale (e.g., 1=not at all to 5= a great deal) to rate a list of items for each question. In addition, background questions were asked about participants' profession, length of experience, type of position, age, gender, and residential location. These background questions allowed us to compare the responses of landscape architects and other professions and to compare study participants from different types of practice, experience, and geographic location.

Data analysis

Categories within the data were identified using principal axis factor analysis with varimax rotation and pairwise deletion of missing data (Norusis, 1993). The criteria used for inclusion of items in a factor category were loadings greater than 0.50 in a category and no dual loadings of greater than 0.50 in two or more categories. Factors were required to have eigenvalues of greater than 1.0. The output of the factor analysis program was used to identify highly coherent and stable categories. In order to enhance internal validity, categories were required to have a Cronbach's coefficient of internal consistency (Chronbach, 1951; Nunnally, 1978), Alpha of at least 0.70 (except as noted). Scales were then constructed using a respondent's average rating of the items that formed the category. Paired sample t-tests and one-way analysis of variance (ANOVA) were used to compare different participant groups.

Results

Experience

In general, the study participants had less experience than we anticipated with ecological restoration. For factors related to ecology and hydrology, such as restoration ecology, stormwater management, wetland mitigation, and stream daylighting, participants had a mid-level of experience or knowledge (mean=2.98). For restoring damaged landscapes, such as mine and landfill reclamation and brownfield rehabilitation; participants, on average, indicated an even lower than mid-level experience (mean=2.34).

Important Goals of Ecological Restoration and Remediation

A central objective of the survey was to understand if landscape architects and other professionals held the same objectives for ecological restoration and remediation as had been described in the literature in the field. Study participants were asked to rate a list of 38 potential goals and objectives for their importance for ecological restoration and remediation. A statistical technique called factor analysis was used to group these items into coherent categories. Eight categories were derived from factor analysis along with several items which did not fit into any category (Table 11.1). According to study participants, the most important goal of ecological restoration was to clean the environment and reduce pollution on land and water (category mean=4.35). Two highly rated items which did not factor on any category were to heal past environmental degradation and reconnect people to the land (mean=4.40 and mean=4.33, respectively). Research in the form of developing new rehabilitation techniques and advancing scientific study of restoration was perceived as the next highest rated goal for ecological restoration (category mean=4.09). Restoring ecology was an equally highly rated category (category mean=3.91) and included items such as improving bio-diversity, restoring native species, removing non-native species, and replicating natural processes. The next highly rated goal (category mean=3.81) had to do with public education and raising environmental awareness about harmful environmental practices, interpreting ecological processes, and instilling a deeper caring for nature. This category also had an aspect of public participation, such as engaging the public in design and management, providing volunteer opportunities and demonstrating stewardship for the land. Creating a place for human needs was the next category in importance (category mean=3.67) and included such items as creating useable open space, provide social benefits, build community, create beautiful places and improve a community's image. It is interesting to note that artistic expression was the lowest rated item in this category. However, cleaning a site for human use did not necessarily mean cleaning up brownfields for new development which received a somewhat low category rating (mean=2.83). While still considered important objectives, highlighting past uses and interpreting historic and cultural artifacts was perceived as of lower importance (category mean=3.43) than many of the other goals.

Table 11.1 Important Goals and Objectives for Ecological Restoration

Categories and Individual Items	Category Mean	Cronbach Alpha
Clean-up the Environment	4.35	0.60
Improve water quality		
Clean-up sites for human use		
Reduce pollutants in land and water		
Research	4.09[a]	0.84
Develop new rehabilitation techniques		
Advance scientific study restoration		
Restore Ecology	3.91[ab]	0.83
Restore native species		
Remove non-native species		
Provide wildlife/aquatic habitat		
Improve biodiversity		
Restore natural hydrology		
Replicate natural processes		
Public Education and Awareness	3.82[bc]	0.91
Educate about harmful environmental practices		
Reveal humans' impact on the land		
Public education		
Engage public in design/management		
Interpret ecological processes		
Demonstrate stormwater management		
Instill a deeper caring for nature		
Demonstrate stewardship for the land		
Provide volunteer opportunities		
Create Places for People	3.67[c]	0.85
Create beautiful places		
Artistic expression		
Create useable open space		
Provide social benefits		
Improve community's image		
Create places for reflection		
Build community		
Highlight Historic and Cultural Uses	3.43	0.74
Interpret historic/cultural artifacts		
Highlight past uses		
Restore for Development	2.83	0.58
Clean-up brownfields for development		
Appear intentionally managed		
Restore to Natural Appearance	2.52	0.69
Restore site to pre-development condition		
To appear natural/untouched		
Single Items:		
Heal past environmental degradation	4.40	
Reconnect people to the land	4.33	
Enhance recreational opportunities	3.52	
Allow nature to take its course	3.44	
Economic development	3.27	

Note: The means represent the average for all the items in a factor. Items were rated on a five-point scale (1 = not at all important to 5 = extremely important). Means with the same superscript are not significantly different at $p<.05$. The items are ordered within each factor from the highest to lowest factor.

In stark contrast to many of the proponents of ecological restoration in other fields, this sample of landscape architects and other professionals did not think ecological restoration should restore a site to pre-development condition nor should the final design appear natural or untouched. This category received the lowest average score of any of the eight categories of objectives (category mean=2.52).

The Role of Landscape Architects in Rehabilitating Degraded Landscapes and Waterbodies

The next survey question asked study participants to describe the role that landscape architects should play in rehabilitating degraded landscapes and water bodies from a list of 11 items ranging from designer and artist to environmental advocate and researcher (Table 11.2). Participants thought the most important role for landscape architects was a designer who provide vision for restoration projects. Participants also strongly indicated that landscape architects should be the team leader in ecological restoration projects. As a team leader, some of the other important roles were as a facilitator and environmental advocate. In the words of one study respondent, landscape architects should "be a voice for nature on design and development team where others are focused on [the] economic aspects of the project." Being an educator about the environment and social advocate were also important roles for landscape architects. The social importance of the profession was eloquently stated in the words of one participant, the landscape architect should be the "restorer and preserver of community and nature." Other roles for the landscape architect received moderate ratings including researcher, negotiate the regulatory process, artist, and technical expert. It is interesting that most participants were able to differentiate the role of designer in a project from that of artist. Landscape architects may respect and support the work of environmental artists in ecological restoration, but they perceive the landscape architects' role as distinct from their allies in the arts. Moreover, landscape architects also acknowledged that one of their least important roles was that of technical expert. This observation makes sense considering that the profession is very broad and may not give one expertise in the sciences, such as hydrology, toxicology, or ecology.

Table 11.2 Role of Landscape Architect in Ecological Restoration

Individual Item	Mean
To provide vision	4.57
Designer	4.56
Team leader	4.20
Facilitator	4.08
Environmental advocate	4.07
Educator	3.92
Social advocate	3.81
Artist	3.56
Negotiate regulatory process	3.48
Researcher	3.47
Technical expert	3.36

Note: Items were rated on a five-point scale (1 = not at all to 5 = extremely well).

In a follow-up question, participants were asked how the role of landscape architects in rehabilitating degraded landscapes and water bodies is different than their role in more traditional "greenfield" development projects on previously undeveloped sites. This open-ended question was answered by about half of the respondents. Reactions were evenly divided with approximately half of these respondents indicating that landscape architects played the same role in restoration projects as "greenfields" development, and the other half indicating that there were indeed differences in the landscape architects' role in degraded landscapes and waterbodies. One of the main differences between greenfields and brownfields projects that survey participants described was the complexity and unknown factors that are typical of degraded landscapes. In the words of one participant, "The root design principles are essentially the same, but health considerations and hazardous materials and contamina[tion] issues make it much more problematic and complex." This scientific complexity of degraded landscapes meant that according to another respondent, "most remediation projects demand a high level of technical expertise that lie outside the traditional role of the landscape architect (e.g., hydrology, geochemistry, civil and chemical engineering, law, and geology)." Another important differences was in the dynamic quality of natural processes. In the words of one participant, it "requires a new mindset on processes over design state...requires scientific commitment to dynamics over visual." Another participant went on to say, "individual expression, artistic style [are] less important" in restoration projects. From this perspective, it would be interesting to know which skills and knowledge study participants thought would be most important for landscape architects to have to be effective in ecological restoration projects.

Participants ratings of 16 types of skill and knowledge were grouped into categories using factor analysis (Table 11.3). While the majority of skills received high ratings, the most important skills included traditional design skills, grading and drainage, and problem solving/ critical thinking (category mean=4.59). The ability to problem solve and think critically was by far the most highly rated item of the list. Project management was considered next highest rated skill (mean=4.23). Another set of important skills (category mean=4.11) involved the natural sciences of stream and wetland ecology, hydrology, soils and geology, along with stormwater management. Other scientific areas also received high ratings, including landscape ecology and botany and plant material knowledge, but did not cluster in any category. An equally highly rated set of skills was understanding the regulatory process and knowledge of water quality standards (category mean=4.06). Research skills were also considered important, but were not the highest rated.

Table 11.3 Important Skills and Knowledge Areas for Landscape Architects to be Effective in Ecological Restoration

Categories and Individual Items	Category Mean	Cronbach Alpha
Traditional Design Skills	**4.59**	**0.67**
Design skills		
Problem solving/critical thinking		
Grading/drainage		
Project Management	**4.23**[ab]	
Natural Sciences	**4.11**[bc]	**0.84**
Stream ecology		
Soils/geology		
Wetland identification		
Hydrology		
Stormwater management		
Knowledge of Regulatory Process and Standards	**4.06**[ac]	**0.78**
Knowledge of water quality standards		
Understanding of regulatory process		
Single Items:		
Landscape ecology	4.45	
Botany/plant materials	4.35	
Public participation	4.15	
Construction technology	4.01	
Research skills	3.93	

Note: The means represent the average for all the items in a factor. Items were rated on a five-point scale (1 = not at all important to 5 = extremely important). Means with the same superscript are not significantly different at $p<.05$. The items are ordered within each factor from the highest to lowest factor.

The study participants were asked to go back to the list of important skills and knowledge areas for ecological restoration and indicate which important skills or knowledge that landscape architects should have, but were currently lacking or inadequately covered in landscape architectural education (Table 11.4). Only about half of the study participants answered this question. Of these respondents, approximately 40% indicated that project management was an important knowledge that was lacking in current landscape architectural programs. Many participants (37%) also indicated that landscape architects were not adequately prepared by their education to facilitate public participation in planning and designing ecological restoration projects. In the words of one participant, "restoration and rehabilitation are often public and large-scale projects. They take a long time and involve many participants. Both persistence and political skills are critical for success." Other important areas of knowledge for ecological restoration that deserved more coverage in landscape architectural education were the sciences of hydrology, stream ecology, landscape ecology, and soils and geology, which were mentioned by 30-40% of respondents respectively. Not surprisingly, traditional skills such as design, grading and drainage, and construction technology were felt to be well covered in current educational programs. Interestingly though, almost 25% of the respondents indicated that problem solving and critical thinking skills could be better covered in landscape architectural education to prepare the profession to be more effective in ecological restoration projects. It appears that there are several key areas that need to be addressed in training landscape architects for a leadership role in healing degraded land and waterbodies, including the natural sciences, project management, public participation and problem solving. It would be interesting to know what other impacts this new field of practice may have on the landscape architectural profession.

The Impact of Ecological Restoration on the Landscape Architecture Profession

Ecological restoration and remediation work may impact the future direction of landscape architecture. In order to understand where these changes might take the profession, study participants were asked to indicate the extent to which they agreed with ten possible changes to the profession that may occur as part of landscape architects' work in this area (Table 11.5). In general, there was a very high level of agreement with the statements (average means were above 4.00 on a 5.00 scale). The three equally highly rated categories that were derived from factor analysis involved office structure and practice, leadership, and job opportunities (category means=4.20, 4.06, and 4.11, respectively).

Table 11.4 Important Skills and Knowledge Areas for Ecological Restoration that are Currently Lacking or Inadequately Covered in Landscape Architectural Education

	Frequency	Valid %
Project management	18	41.9
Hydrology	18	41.9
Stream ecology	16	37.2
Public participation	16	37.2
Landscape ecology	15	34.5
Stormwater management	14	32.6
Soils/geology	13	30.2
Botany/plant materials	11	25.6
Wetland identification	11	25.6
Understanding of regulatory process	11	25.6
Problem solving/critical thinking	10	23.3
Knowledge of water quality standards	8	18.6
Research skills	8	18.6
Construction technology	5	11.6
Design skills	2	4.7
Grading/drainage	2	4.7

Note: Valid Percent (N=43)

With regards to professional practice, study participants thought that involvement in ecological restoration would lead to more multi-disciplinary offices and a need for more continuing education courses. This would also include more long-term monitoring of projects. With regards to leadership, the second category, participants thought participation in ecological restoration may lead to a greater leadership role for landscape architects, greater visibility for the profession, and the opportunity to get involved earlier in the planning process. With regards to job opportunities, the third category, study participants predicted that ecological restoration work would lead to a greater number of projects and job opportunities for landscape architects including more research-based projects. Two items which did not fall into any category, were the prediction that ecological restoration would lead to more specialization in the profession and more involvement in regulatory initiatives, which was perceived as the least likely impact on the profession.

Table 11.5 Impact of Ecological Restoration on Landscape Architecture Profession

Categories and Individual Items	Category Mean	Cronbach Alpha
Multi-disciplinary	**4.20**[ab]	**0.75**
Increased need for more cont. ed. classes		
More long-term monitoring of projects		
More multi-disciplinary offices		
Job Opportunities	**4.11**[ac]	**0.84**
Greater number of projects/job opportunities		
More research		
Increased Leadership and Visibility	**4.06**[bc]	**0.77**
Greater visibility		
Greater leadership role		
Involved earlier in the process		
Individual Items:		
More specialization within the profession	3.82	
More involved in regulatory initiatives	3.67	

Note: The means represent the average for all the items in a factor. Items were rated on a five-point scale (1 = not at all to 5 = a great deal). Means with the same superscript are not significantly different at $p<.05$. The items are ordered within each factor from the highest to lowest factor.

Participants were then asked to answer and open-ended question about how the current trend in ecological restoration work has affected the direction of their own career or office. Over half of the respondents (55%) indicated that ecological restoration had made some impact on their career or office. Many of the students indicated that ecological restoration had made a strong impact in their decision to become landscape architects. Furthermore, they see some impacts of ecological restoration on education programs with "new courses being offered focusing on sustainable or ecological design." For the professionals in the study, ecological restoration has "open[ed] new perspectives and opportunities." In the words of another, it has "provided wonderful opportunity to provide skills in environmental and social benefits. [It] provides 'sense of purpose' for my career." Collaboration and multi-discipline teams were other themes in the comments. One professional indicated changes such as "more collaborations and the need to study a greater breadth of material" and another described the need to "manage a highly diverse multidisciplinary team." The scope of landscape architectural practice has also been affected, "It has caused us to look beyond the narrow scope of our work. We are more aware of the needs of nature and maintaining or restoring. We also must look to future needs of the community." However, there was also some frustration expressed by young professionals, "My interest in ecological and cultural questions is what led me into this profession. Unfortunately, in my limited professional experience so far, I have found virtually no interest in this area among my co-workers." For those who have found the right office or ecological restoration project, the impacts have been profound. As one professional exclaims, "My current project is ecological restoration and I gotta tell ya [sic]; it feels like the right thing to do!" These enthusiastic professionals see "a huge opportunity for landscape architects to take a leadership role in repairing damaged landscapes."

Group Differences

Previous research has shown that environmental experience is an important factor differentiating people's support for ecological restoration projects and landscaping with native plants (Ryan, 2000; Nassauer, 1995). In the current study, environmental experience was measured in several ways. Participants were asked to assess their knowledge and experience with respect to different aspects of ecological restoration. This question revealed two different types of experience, experience with ecology and hydrology and experience restoring degraded landscapes. Participants were also asked to indicate their professional affiliation (i.e., landscape architect, engineer, etc.) and their level of experience in their current profession. Participants in these different groups were then compared using statistical techniques (i.e., t-tests and analysis of variance) about their responses to the four sets of survey questions: the most important goals of restoration, the role of landscape architects in ecological restoration, the skills that are important for landscape architects to have to be effective in ecological restoration, and the impacts that restoration will have on the profession. Comparisons were also made with regard to residential location (i.e., domestic versus international participants), and age.

In general, there was considerable agreement in response to the survey questions between those who had little experience in ecological restoration and those who were very experienced. The most interesting difference was in regards to the goals of ecological restoration. Those participants who indicated a high level of experience restoring damaged landscapes (i.e., brownfield rehabilitation, mine and landfill reclamation) were significantly less likely to indicate that a site should be restored to a natural appearance (mean=2.27) than did those with less experience with restoring damaged landscapes (mean=2.89, t=-2.73, p<.01, d.f.=87). However, those participants with more experience rehabilitating degraded landscapes were significantly more likely to think these sites should be restored for development and appear intentionally managed (mean=3.01) than did those with less experience (mean=2.48, $t=2.79$, $p<.01$, d.f.=85). These differences could reflect the fact that those who have dealt with the complexities and serious degradation of brownfields, landfills, and mines realize the difficulty in restoring these sites to any semblance of a natural, pre-settlement condition and thus, restoring these degraded sites for more intense human use may be seen as a more practical and efficient use of land.

Professional affiliation was the most significant differentiating factor among survey participants' responses. Landscape architects were significantly less likely to think ecological restoration should restore a site to a natural, pre-settlement appearance (mean=2.26) than did other professionals in the sample (mean=2.93, $t=-3.05$, $p<.005$, d.f.=86). However, landscape architects were significantly *more* likely to think that ecological restoration should highlight a site's historical and cultural features and uses (mean=3.59) than did other study participants (mean=3.22, $t=2.00$, $p<.05$, d.f.=86). Landscape architects also viewed their role in ecological restoration differently than did survey participants from other disciplines. By a huge margin, the landscape architects saw their role as team leader in ecological restoration projects (mean=4.55) than did other participants (mean=3.74, $t=4.35$, $p<.000$, d.f.=85). Landscape architects were also significantly more likely to perceive the role of landscape architect as designer and facilitator on restoration projects than did other professionals. It was interesting that there was more agreement between landscape architects and other professionals on the impacts that ecological restoration would have on the profession and the necessary skills that landscape architects should have to be effective in restoration work.

Length of experience in one's profession also had a minor influence on perceptions of the landscape architects' role in ecological restoration. Those with less than 10 years experience in their respective fields were significantly more likely to perceive the role of landscape architect in ecological restoration as designer (mean=4.74) than did those with 10 or more years experience (mean=4.38, $t=-2.27$, $p<.05$, d.f.=74). These same differences were also found when comparing younger and older participants, since experience in one's profession is often related to age.

Finally, there were several significant differences between international participants and those from the United States about the role of landscape architects in ecological restoration. International participants, who were only a small sub-set of the sample, were significantly less likely to perceive landscape architects' role as one of designer in ecological restoration projects (mean=3.91, $F=5.73$, $p<.05$, d.f.=2,83). Not surprisingly, they were also less likely to put an emphasis on traditional design skills as being important for landscape architects' effectiveness in ecological restoration. These responses may reflect the different ways in which landscape architecture is practiced domestically versus internationally.

Insights for Environmental Restoration and Design Education and Practice

It is important to understand the motivations behind those engaging in ecological restoration (see Chapter 3 by L. and S. Conn and Chapter 4 by D. Kidner). Through a survey of landscape architects, environmental designers, and other restorationists, this study revealed new insights into the perspectives of each group. According to the current study, the most important priority for ecological restoration projects is to heal past environmental degradation and clean up the environment by reducing pollution on land and water. This perspective was widely shared by the many professions represented in the study. Differences between landscape architects and other professionals engaged in ecological restoration appeared when looking at the ultimate end-use for a restored site. From the perspective of landscape architects, the goal of ecological restoration is not to restore sites to some idealized, pre-settlement condition, but to clean up polluted land and waterbodies for the benefit of both humans and wildlife. This is in marked contrast to the viewpoints espoused by many restoration ecologists who think the primary goal of ecological restoration is to benefit native plant and animal species by recreating environments where the signs of human impact on the land have been erased (Throop 1997, 2000). However, this is not to say that restoring ecology was not strongly supported by landscape architects and other design professionals in this study— it was. Yet, environmental designers, particularly landscape architects, appeared to take a more balanced approach to ecological restoration when it comes to balancing human needs. A strong emphasis on reconnecting people to the land has also been proposed by many restoration ecologists (Jordan 1989; see also Chapter 6 by S. Mills, Chapter 2 by M. Dannenhauer and Chapter 1 by W. Jordan and A. Turner). In general, reconnecting people to the land from the restoration ecologists' perspective has been through volunteer stewardship programs that emphasize restoring native ecosystems (Ryan et al. 2001; see Chapter 5 by Light). However, from the landscape architect's perspective, reconnecting people to the land means not only raising public awareness through interpreting restored landscapes, it also means reconnecting people to a site's historic and cultural past, as discussed in Chapter 10 by Mozingo and Chapter 7 by E.

Spelman. Several projects presented at the " Brown Fields, Grey Waters" Conference, such as the Westergasfabriek site in Amsterdam and the work of Ukeles at Fresh Kills landfill in New York City (O'Connell 2001) exemplify this more holistic approach to restoration, as is discussed in Chapter 9 by T. Collins and Chapter 8 by J. Brown.

The results of this study lend support to those who have argued that ecological restoration must first and foremost create cleaner, safer environments for people (Lodiek 1994; Steingraber and Hill, 2002). From this perspective, it is not surprising that those study participants with more experience in brownfields and landfill reclamation put a significantly stronger emphasis on cleaning up degraded landscapes for human use including new development. By supporting human's need for clean living, working, and recreational environments, restoration ecology can garner wide spread support for projects that must compete for diminishing private and public funds for environmental protection. Furthermore, cleaning the environment for people benefits the native plant and animal species who rely on clean soil, air, and water for their survival as well.

Implications for the profession and education

The study revealed that ecological restoration has had and will continue to have profound impacts on the landscape architectural profession and education. The scientific complexity and unknowns that are inherent in degraded land and water bodies require a multi-disciplinary approach to ecological restoration and landscape architectural practice. The scientist, researcher, and environmental policy analyst need to be woven into the traditional design team of architect, landscape architect, artist, and engineer. Successful collaboration between professions requires some base knowledge and respect for the perspective of other professions and disciplines. For landscape architectural education, this means more emphasis on the natural sciences, as well as skills in project management and public participation. The importance of learning to work in interdisciplinary teams has been proposed by landscape architects and ecologists who are interested in integrating ecology into design education (Adams 2002, Nassauer 2002, Karr 2002). The role of the landscape architect may not be the technical expert on hydrology or phyto-remediation, but he or she must be conversant in the language of the other respective disciplines. One must be able to know the questions to ask the other members of a project team and know the parameters and world-view that guides the other respective professions. This widespread knowledge about related knowledge areas in ecological restoration is key if landscape architects are to take the leadership role in ecological restoration that they aspire to in this survey. The results of the survey also suggest that currently, other related professions are much less likely to perceive landscape architects as the team leader for ecological restoration projects. The landscape architecture profession is at a key decision point about taking the lead in ecological design (Adams 2002, Johnson and Hill 2002). Landscape architectural professionals and educators will need to improve the profession's skills in multi-disciplinary collaboration, project management, public participation, the natural sciences, and environmental regulation in order to become leaders in ecological restoration.

Conclusion

Ecological restoration is just one piece of a larger world view about human's relationship to the environment. The first premise of a sustainable or environmentally sensitive design and planning is to minimize harm to the natural environment which in many cases means steering development away from sensitive areas. The larger goal must be to respect existing healthy land and waterbodies in the environmental designers' efforts to accommodate increasing human demand for scarce land and water resources. Ecological restoration has the ability to heal previous environmental damage, but it should not be at the expense of more pristine environments. Pennsylvania's land-recycling approach to its many brownfields sites is one example of steering new development to land that has already been impacted by human use (Pennsylvania Department of Environmental Protection 2000).

The greater challenge for the environmental design professions is to better integrate ecology into all aspects of practice and education. (See Johnson and Hill 2002 for key essays on this topic.) This environmental ethic for professional practice is summarized by one of our study participants, "I believe ecological restoration of native landscape systems is one facet of ecologically-based design. We must resolve the conflict between development and preservation by making *every place* more beautiful and healthy through the employment of ecologically-based design and construction technologies. Buildings, infrastructure, and landscapes can all be compatible with ecologically sustainable places by placing primary emphasis on the ecological performance of these things."

Conclusion

Landscapes and Mindscapes of Restoration

Robert France

Restoration design is the process by which participants creatively develop physical and conceptual relationships to engage repaired nature through the architectural transformation of their inhabited ecological space as well as their internal consciousness.

Premise I. Restoration Design is Landscape Architecture

"The idea of *nature by design* is anathema to most restorationists, ecologists, and environmentalists."

—Higgs, E., *Nature by Design: People, Natural Process, and Ecological Restoration,* 2003

Existing Limitation

Restoration ecologists continue to refuse to admit that what they are doing is really a specialized form of landscape design and social revitalization, instead deluding themselves into believing in the naturalness of their actions in a quixotic quest for that elusive Sangrael of precise replication. And in so doing, these self-imagined knight-errants of ecological purity raise the dire of the, in their minds, arm-waving and finger-pointing theoreticians who just won't leave them alone to do their important jobs. This latter group, in turn, is bothered by both the metaphysics and the implicit hubris behind the chasing of the restoration chimera. For there is no denying that restoration, at least that as heralded by many ecologists, is like venturing through a wormhole in a bad science fiction novel or movie; i.e. one must predict the past in order to restore the future. It simply doesn't make sense, and to worry about the paradox too much is to hurt one's head. And then, to continue with the time travel analogy, just whose idea of which time should we be shooting for in our designs? Indeed to some theoreticians, the above analogy is not far off the mark as they firmly believe that restoration is often more science fiction than it is science fact. One also gets the feeling that the restoration practitioners at times don't really care one way or another, regarding the whole debate as little more than a silly distraction from their crucial work (if they're feeling magnanimous) or as a wasteful diversion of fiddling while the world is burning (if they're feeling angry). All these dilemmas also play out in the minds of individuals as well. For example, as a scientist, one who has at times conducted field research in some of the most remote areas of wilderness on the globe, part of me is sympathetic to the concerns of the ecology camp. However, as a professor in a design school in which cultural history and art are expressed and celebrated with just as much vigor as is the desire for biological diversity, I find myself frustrated by the rose-tinted glasses that seem to cloud the vision of many of my ecological brethren engaged in restoration work.

Future Promise

Environmental theoretician E. Higgs concluded his informative book about the developmental theory of restoration with a chapter that gave rise to the title of his book *Nature by Design,* playfully tweaking the title of landscape architect Ian McHarg's seminal book from the 1960s, *Design With Nature.* What is most surprising is that if ecologists made it that far into Higgs' book, they might have been taken back by this conclusion about restoration being design. Indeed, though pitched as a controversial revelation by Higgs in his book (written with an audience of restoration ecologists in mind), landscape architects, steeped as they are in a near idolization of Olmsted and his brilliant restoration work on the famous Back Bay Fens of Boston of more than a century ago, merely shrug knowingly, saying that that is what restoration has been about all along. To imagine that restoration is anything more noble and "natural" than simply rewriting over and adding to the cumulative landscape palimpsest is, argue landscape architects, missing the entire point that what we are dealing with in "restored" sites is creating cultural entities. After all, "landscapes" by their very definition embrace the inherent teleology embedded in the words "nature" and "design."

Book Contribution

Chapters by L. Mozingo and R. Ryan in the present book demonstrate that restoration design is very much its own entity, separate in spirit and purpose from, and more holistic in both its conception and execution than, restoration ecology. Indeed, though elements of the latter are subsumed within the former, restoration designers are also much more honest in their acknowledgement of the role of humans in shaping the natural world. As a form of landscape architecture, restoration design is one of the most integrative of all environmental disciplines and has a rich history that extends back in time well before Aldo Leopold's ecological experiments in the mid-1930s. Indeed, it may be argued that the ostensive birth of restoration design is really also that of "modern" landscape architecture.

Premise 2. Restoration Design is Creation

"Science is the art of knowing; art is the science of feeling. We must know in order to do; but we must feel in order to know what to do."

—Cauldell, C. *In* Eckbo, G. *Landscape for Living*, 1950

Existing Limitation

Until recently restoration has been the near exclusive perview of scientists and has mostly ignored the important work that environmental artists can bring to any project to increase its public visibility and acceptability. The absence of public understanding about restoration as practiced by scientists has had the result of often marginalizing such projects. And the zeal for re-creating "naturalness" in restored landscapes by some ecologists has also fed the public's ignorance about the deliberate and technically challenging interventions being implemented. As a result, restoration is thought of as something that takes place in remote regions far from cities. This occurs due to the reticence of the great majority of ecologists to have anything to do with urban landscapes; i.e. cities after all are palaces of culture, *not* places of nature. But as H. Dreiseitl asks, "Why do we have a civilization where our dreams are so far away? Why can't we work in a way that our living space have a quality that this is our new nature – our urban nature?"

Future Promise

Whereas ecologists flee cities, artists thrive there and thus have much to contribute to fostering restoration design. Artists function as facilitators and translators, acting as important go-betweens in restoration design, transferring the science and engineering information to the community in ways that mobilize action. Many eco-artists may not attempt to treat and fix nature but rather believe it enough to act as radar systems in drawing public attention to problems. In other cases, however, some environmental artists are not content to assume a passive role of merely commenting upon the world but pursue various inventive strategies to physically transform their surroundings. As W. Jordan writes in his book about restoration and community: "It is important to keep in mind that restoration was an artistic enterprise well before it became a scientific one."

Book Contribution

Chapters by J. Brown and T. Collins in the present book show that restoration design is neither pure science nor pure art, but a creative blending of the two, an odd hybrid that at times is seemingly straightforward but may often be a perplexing paradox of intent and execution. Restoration designers are those individuals who effortlessly either integrate these two important spheres of human creativity themselves or recognize the importance of assembling inter-disciplinary teams that collectively can provide such integration. Restoration design is in the end about thinking with the heart, and feeling with the mind, a soulful union of respect and remembrance with that of hopeful anticipation.

Premise 3. Restoration Design is Ecopsychology

"The problem of rescuing water from death must therefore be solved inside ourselves before we can solve it in the external world. When we have transformed the inner sense, the outer one will be restored to order."

—Schwenk, T., *Water, the Element of Life*, 1989

Existing Limitation

Reviewing the literature on environmental restoration, both theoretical books like the present one as well as technical papers by field practitioners, one cannot help but notice the disjuncture that seems to exist. It is almost as if there are two solitudes, each group operating within such different world views that when they do occasionally intersect, the feeling that emerges is more one of surprise than of expectation. Ecological restoration for the most part remains a nuts-and-bolts physical activity, a battle fought in the denuded trenches, the luxury of reflective thought (sociological, historical, philosophical, metaphysical and psychological) being something engaged in as an afterthought. There is little doubt that the current preoccupation of our educational system with technical specialization inadequately trains scientists to successfully engage in such an all encompassing pursuit as restoration. It is the rare practitioner, it seems, who pauses long enough to conceptually attempt to situate her or his restoration projects within a template of environmental action (deep reform, partial reform, incremental advances, holding the line, or slowing the rate of retreat) and also how that placement in turn is effected by, and effects, the internal landscape. In short, most who practice restoration treat it as an end in itself rather than as a means towards something that can benefit the internal world as much as the outer world.

Future Promise

The working precept of ecopsychology is based on the supposition that it is impossible to have healthy people on a sick planet, for the well-being of one influences that of the other. Therapists have recently begun to recognize that there is never a sharp divide between the inner and outer worlds, but instead a continuum extending from the scale of the planet to that of the person. In this sense, the world can be regarded as an extension of ourselves, with a synergistic interplay between the two. Ecopsychology concerns itself with exploring the motivations, yearnings, needs, and ideals that shape and structure our lives within the environment, focusing on strengthening or even reawakening the reciprocal relationship. Given the possibility that restoring nature can come from within, ecopsychology examines whether it is possible to improve the state of the internal landscape by prescribing nature for healing, the hope being that one could then reciprocate and work toward restoring environmental health.

Book Contribution

Chapters by L. and S. Conn and by D. Kidner in the present book articulate both the ensuing benefits and the implicit dangers that can develop from a psychological conceptualization of restoration design. A restoration designer with an inherent process-minded, Gandhian sensibility coupled with a physical imperative to act upon that sensibility knows that it is really the means that matter the most, not the ends, and that if attention and thought are paid toward the former, the latter will evolve appropriately as a matter of due course.

Premise 4. Restoration Design is Mutualism

"The path to the great remembering is through the healing of land conservation and the healing of ourselves, through a million different ways to show our forbearance to reconnect with the life that is around us."

—Forbes, P., *The Great Remembering: Further Thoughts on Land, Soul, and Society*, 2001

Existing Limitation

It is entirely possible in modern western society to live essentially completely outside of nature. The domestication of our modern lives has left us sensually crippled, and having lost our relationship with the natural world, we have begun, collectively, to suffer from a form of neurasthenia due to the severance of that relationship. "Contact" for many now means nothing more than having constant access to a cell-phone. We therefore desperately need to rediscover models and metaphors of instruction about how to re-experience and re-connect to the world about us. And it is here where many have suggested that ecological restoration can play an important role in fostering such a renewed relationship. Although sounding good on paper or in a lecture, realistically the problem remains, however, that due to the technical preoccupation with ends, adjudication of the success of restoration continues to remain largely product-, not process-, based. And as such, many practitioners believe that in order to ecologically "restore" a location to some realistic semblance of its past condition, a certain level of professionalism needs to be present; i.e. restoration ecology requires those most knowledgeable about ecology running the show.

Future Promise

Whether considered on the scale of the northern boreal forest with our tacit societal endorsement of fire suppression activities there or on the scale of our own private backyard gardens, we are all landscape designers. To ignore this and to posit environmental restoration as being somehow distinct and ecologically "pure" and in need of a professional priesthood to actualize it, is to miss the point. As has been advanced before, restoration, if approached in an egalitarian way, offers one of the best means in which to reconnect people to the world in a healthy relationship. Direct public engagement in the physical act of restoration is therefore the route to success. The key to the process is the fact that through restoration people develop a way of increasing the bonds of care with which they experience the world. And the way to bring about such a deep immersion is through focusing on relationship building. The goal here is to engage restoration as an effective relationship between the restorer and that which is restored. Not an "efficient," but rather an "effective" relationship. The distinction is an important one, for whereas the former is that which professionals can bring to a project, it is the latter which is in the end that measure by which we all judge the successes of our own inter-related lives.

Book Contribution

Chapters by A. Light, S. Mills and E. Spleman in the present book show how restoration design can engage lay participants in landscapes of memory where their thoughts and actions are anchored to and guided by the accumulated wealth and wisdom of history, both cultural and ecological. In this respect, Marcel Proust would have been a good restorationist, Robert Graves not so. For above all else, restoration design is about the remembrance of, not the bidding goodbye to, that which has occurred in the past. Restoration design is thus a way to unlock the eco-cultural memories housed within our collective imagination and experience. Restoration design is concerned with the process as much as it is with the product of reparation. By instilling a sense of community and of shared purpose, such mindful restoration establishes effective relationships of reciprocity where the landscape influences the inscape and vice versa; where the healer becomes the healed. Restoration design is an eco-humanist, ecumenical and often non-professional public endeavor in which conscious evolution is the final goal, not merely the return of various missing species or former ecological states. By reinhabiting culture within a rehabilitated nature, such restoration design becomes the ultimate form of ecological education, and thus a true quasi-spiritual undertaking.

Premise 5. Restoration Design is Optimism

"We have much healing and recovery to do, but we have an unbroken spirit."

—Herbert, B. and K. J. Anderson, *Dune: The Butlerian Jihad,* 2002

Existing Limitation

Restoration is a positively minded way to look upon the world rather than moping about the various degrees of nature lost, damaged or destroyed. As currently practiced, however, the track record of restoration in rebuilding hopeful human communities in affiliation with the reconstructed natural communities is not what it could be in this regard. Restoration ecology is focused on the paradigm of reassembling the broken bits and pieces of nature in attempt to set the clock back to some pre-stressed state. It is most frequently approached as a scientific project, infused with Baconian rationalism. Little room exists in such a mindset for mystery, mythology or metaphysics. As a result, there often seems to be a distance that limits the wider acceptance of restoration ecology among the public due to the reticence of many scientists to leave behind, even if only temporarily, their comfortable world of analytical reductionism in order to venture into the realm of subjective holism. However, in any individual's life, the most important decisions are almost always those deeply routed in subjectivity. The frequent absence of a way in which to mourn the loss or to celebrate the rebirth of original and restored ecosystems through some form of ceremony has meant that the public has often missed an opportunity to become more involved in the ultimately hopeful enterprise of restoration.

Future Promise

Community develops through shared participation in common tasks. And one of the most important of such tasks that can bind people together, whether it be in collective happiness or in mourning, is ceremony. Ceremonial performance rituals are those touchstones through which the imprint of time and space are inscribed upon our lives and through which, as a result, we situate ourselves and structure our lives in an otherwise amorphous chaos. The communal sharing of guilt or grief as well as the celebration of genesis or growth, can all be aspects of restoration that strengthens bonds of affection and connectivity to the nonhuman world. In his book *The Sunflower Forest: Ecological Restoration and the New Communion With Nature,* W. Jordan argues that engaging in such invented ceremonies during the process of restoration will instill both value and awareness of the projects.

Book Contribution

Chapters by W. Jordan and A. Turner and by Mark Dannenhauer in the present book illustrate a richness of celebratory approaches involving ritual and performance through which to practice community-based restoration design. The strongest piece of commonality that cuts across and characterizes all opinions and actions of those involved with restoration design is an overwhelming feeling of both the magnitude and the importance of the work, as well as of the unbridled hope brought about through participating in these activities. Although we may very well live in a broken "world of wounds," all restoration designers posses an "unbroken spirit" and remain confident through their thoughts and actions that the "end of nature" is not nigh.

Epilogue

Swamped! A Tale of Two Restorations: Part II: The View To Abroad

Robert France

"Ever the river has risen and brought us the flood,
the mayfly floating on the water.
On the face of the sun its countenance gazes,
then all of a sudden noting is there."

—*Epic of Gilgamesh*, 1200 BCE, *In* France, R.L. (Ed.) *Wetlands of Mass Destruction: Ancient Presage for Contemporary Ecocide in Southern Iraq*, 2007.

Refocusing from staring out the window and thinking about the former Great Swamp, I glance again at the pile of international newspaper clippings on restoration scattered about my desk and cannot but help to remember the demonstration on the Boston Common in early March 2003. There among the carnival-like crowd of thousands of peace protestors with their colorful signs decrying America's immanent preemptive strike against Iraq, was a group of several dozen citizens vociferously arguing for the need to go to war against

Iraq. And there, off to the side as if even he was somewhat embarrassed by the jingoistic calls, flags and "support our troops" signage being waved around by his colleagues, was an individual whose own sign simply read: "Invade Iraq to Save and Restore the Garden of Eden."

The Mesopotamian marshlands of southern Iraq, originally occupying an expansive area of over 20,000-square kilometers, were once one of the most important freshwater wetland systems in the world. The marshlands were the permanent habitat for millions of birds and a flyway stopover for billions more migrating between Africa and Siberia. They also functioned as a nursery habitat for millions of migrating coastal fishes and shrimp from the Persian Gulf. All this life brought people to the area as well. Since ancient Sumer the indigenous dwellers of the marshlands have lived on islands constructed of reed and mud, depending on fishing, hunting and limited forms of agriculture, and developing a unique culture steeped in tradition and the sustainable use of the wetlands.

After 5,000 years, all this came to an abrupt end in the 1990s. Large upstream diversions (some occurring in neighboring countries) of the once sustaining flows from the Tigris and Euphrates rivers began a process of desiccation exacerbated by repercussions from the failed Shi'a uprising at the end of the first Gulf War. Fleeing into the marshlands, the rebels and their temporary landscape refuge became the target of the Baghdad regime. Saddam's military invaded the marshlands killing thousands of marsh Arabs and driving many more across the border into Iran to begin their lives there as refugees. A systematic program was initiated to erase the wetlands from existence. Over 300 kilometers of huge man-made rivers with self-aggrandizing names were constructed to effectively drain vast reaches of the marshlands. Further water was channeled away in networks of drainage ditches and pumping stations to irrigate other areas for unsustainable wheat production. And enormous embankments sealed off other portions of the marshlands, the evaporating water producing large salt-pans of ruined and sun-baked soil. The scale of this ecocide is almost beyond comprehension and modern precedence, being likened to the drying up of the Aral Sea or the deforestation of the Amazon. Less than seven percent of the original marshlands remained.

Some Biblical historians believe the one-time lush and bountiful desert wetland oasis of the Mesopotamian marshlands to have been the inspiration for the Garden of Eden myth. This is a glaring aberration, however, with respect to how humankind, no matter where in the world or when in history, generally viewed wetlands. For no single type of ecosystem has born the brunt of human antipathy more than have wetlands. Neither real land nor real water but a messy and indeterminate intermingling of the two, for the most part we either fear or are repulsed by wild wetlands because we simply don't know where to stand (both physically and conceptually) with respect to them. When historically we have admired wetlands, it has had to have been on our own terms as little controlled bits of the real thing presented as manicured ornamental gardens as in Suzchou or Kyoto. And that is why the apostasy of Thoreau is so singular. Our negative view of wetlands as being harbingers of criminals or sources of pestilence continues to fill our language with pejorative terms; i.e. being "swamped" in work, "bogged down" in details, or "mired" in problems are certainly not admirable states to find oneself in. All the more remarkable then to have seen this single person on the Boston Common, surrounded by thousands of peace protestors (many of whom who would no doubt proudly refer to themselves as "environmentalists"), arguing instead for the need to go to war and justifying his belief in the righteousness of the act based on the damage already wrought upon the marshlands by Saddam, the need to prevent that regime from continuing his actions, as well as the need to salvage that which has already been lost. It is of course not new that nature can be used as a political football nor that ecological restoration can be conceptually complicated, but the extent to which the very idea and the fate of the Mesopotamian marshlands played in both the buildup to the war and the consequent rebuilding of postwar Iraq may be unprecedented in contemporary restoration theory. The reason may stem from the moniker applied to the marshlands as being the putative site of the Biblical "Garden of Eden."

Despite the rhetoric trying to convince people otherwise, there is little doubt that the assault against the "axis of evil" took on the demeanor of a modern crusade. In early 2003 as the rush to war was in full swing, I was in Syria visiting the wonderfully preserved and restored crusader fortresses and reading many texts on the subject. A few months later, watching broadcasts of American soldiers crouching in the sand holding their rifles upright beside them as they received blessing from a Christian pastor, I couldn't help but see the parallel to the paintings of eleventh-century crusaders kneeling beside their swords pointed downwards into the sand, symbolizing crosses. And if I had any doubts at all about the parallelism, it was always brought back again and again by the eerie coincidence in having the leader of the new crusader forces being named General *Franks*! Upon questioning, the young man standing on the Boston Common turned out to have just such a quasi-religious mindset causing him to want to invade Iraq in order to protect *our* Garden of Eden from being despoiled by *those* Moslem infidels.

Merely preserving the few remaining parts of the mythical wetlands was not enough however. It was necessary, many believed at the time, for us to go into Iraq and to actually *restore* Eden to its former glory. Indeed, with the possible exception of ecological restoration plans for the Everglades and for the Venice lagoon, no such potential project has generated the same extensive press coverage as that for restoring the marshlands of Iraq. Again and again the coverage reiterated what an environmental disaster Saddam had inflicted upon the world, the implicit message being left that it was someone's duty to stop him and to heal the wrong. One could almost imagine a new battle cry forming amongst our troops: "Resurrect Eden Again – Deus Lo Volt!"

Although to be sure, there are few modern parallels to the purposeful zeal and successful implementation with which the Baghdad regime attacked and nearly completely obliterated the marshlands and marsh Arabs of southern Iraq, Saddam is certainly not alone in his enthusiasm for the destruction of such precious wetland jewels in arid regions. The wet-meadows of ancient Palmyra in Syria have been so hydrologically manipulated as to be almost unrecognizable as wetlands today. The once extensive marshes of the Azraq Oasis in Jordan have been essentially sucked dry, the water piped away to feed the growing metropolis of Amman. And arguably the most important wetland of them all in the Middle East (at least in terms of bird migration), the Lake Hula Swamp in Israel, which once was of a size comparable to that of the downstream Sea of Galilee/Lake Kinneret, was nearly completely drained and the indigenous marsh Arabs driven out during the rush for agricultural self-sufficiency as the new country developed in the mid-1950s. Today there are interpretive education centers in the vestigial pieces of wetlands remaining in Jordan and Israel, and recently other areas of former Lake Hula have been successfully restored for ecotourism and water treatment purposes.

For some, ecological restoration carries with it elements and burdens of remorse, shame, guilt, penance, and restitution. Therefore, in an ironic twist, some of those most ardently protesting the invasion of Iraq found themselves shoulder-to-shoulder with supporters of the war, united in their mutual desire to restore the Mesopotamian marshlands, albeit for very different reasons. By being unable to halt what in their mind was an egregiously immoral war, this group of individuals now wished to make amends and believed that the best way to offer compensation for the imagined sins of their own country lay in the ecological and cultural restoration of the marshlands and their innocent human dwellers. Indeed, there could be no better project imagined anywhere for restorationists to become involved in; i.e. given that ecological restoration has been likened to being a form of gardening, what could be better than to restore the original Biblical Garden from which we sprang? As well, gardening is a known therapeutic activity benefiting the gardener as much as the land being gardened. Healing the wetlands of southern Iraq, the pacifists believed, was a concrete way of healing not only a war-torn country that bore the brunt of an unjust invasion but also a war-mongering country embarking on a troubling new world order of military preemptiveness.

There is a powerful forward momentum to begin the restoration process in southern Iraq. Landsat images through time provide an alarming chronicle of the extent of the devastation of the marshlands. Part of the restoration process has already begun with the returning marsh Arabs opening up spillways and reflooding areas of former marsh. A danger exists, however, in that the returned water could mobilize the salt-laden soil and produce further harm to those few portions of the freshwater wetland that somehow miraculously managed to survive the widespread desiccation. To this end, a comprehensive soil and water monitoring program has been deemed necessary in order to guide the restorative efforts. One of the first interventions recommended by the reconnaissance team would be directed at reflooding drained areas near surviving marshes to establish nurseries for producing seeds and native species for replanting devastated areas elsewhere in the region. Wildlife and fish restocking programs are also being planned for the most suitable sites selected for pilot projects.

What makes this particular restoration project so important, however, is the degree of attention being paid to the cultural aspects of restoring the relationships of the marsh Arabs to their environment, not just the physical reparation of the wetlands themselves. The challenge is to find ways to allow for traditional lifestyles to be preserved at the same time as being integrated into the modern world. Immediate target objectives are to design and develop decentralized sewage treatment systems that use constructed wetlands and other phytoremediation techniques of ecological engineering to improve the sanitary conditions of the marsh dwellers. Later plans call for international teams of engineers, ecologists, and ethnologists to blend their collective expertise towards developing a sustainable and thriving cultural and ecological landscape. Elements such as transportation, power, communication, fisheries, and agriculture will be addressed and planned for minimal impact on the restored wetlands. Another major goal is restoring public health and economic well-being.

In the end, ecocultural restoration is about balancing human and environmental health. With respect to the Iraqi wetlands, one of the biggest challenges will be the psychological restoration of the marsh Arabs, a process that cannot be imposed from the outside but must arise from within. In this respect, the job becomes just as difficult as it is important. Restoring devastated physical landscapes—reassembling the broken bits and pieces of nature—will always be the easier task compared to restoring degraded psychological landscapes—the hopes, dreams and aspirations of, in this case, a downtrodden people who have suffered more than a decade of abuse and trauma and a near-complete severance from their ecological home. Achieving the goals of such deep and comprehensive restoration will not come easy but will be possible provided all participants dream large, think hard, and act imaginatively.

Appendix

The Muddy, Messy Means and Mores of "Restoring" a Broken World: A Literature Review

A. Restoration as a Positive and Beneficial Enterprise

- Restoration can be likened to a form of gardening and thus represents a healthy, mutually beneficial relationship with nature
- By attempting to reconstruct broken ecosystems, restorationists enter into a close relationship with nature and learn to more fully understand ecosystems; i.e. not only is nature returned but participants are returned to nature as well
- Restoration implies a loss of innocence in acknowledging the often deleterious influence of humans upon the environment, and is therefore a form of restitution or penance for the implications (i.e. "sins") of that influence
- Restoration is an expressive act, a form of ritual based upon renewal of both environmental and human health
- Because comprehensive restoration is about restoring a relationship with nature instead of merely the structure of nature itself (i.e. restoring the culture of nature rather than simply nature), there will always be positive values in the process of restoration even if the product of these efforts do not exactly replicate the natural elements lost; i.e. the goal of restoration should always be process motivated rather than product motivated
- "Benevolent" restoration should not be confused with "malicious" restoration or development mitigation as the latter can used as an excuse for perpetrating continued damage to nature
- Even if the end result may be an artifact of the real nature lost, engaging in restoration can be beneficial in releasing nature from the human shackles previously placed upon it
- Restoration accepts human participation in a common future with nature and thus gives rise to a new kind of environmental ethic
- Restoration is a constructive dialogue with, and a gift back to, nature in compensation for what it has given us as well as for what we have taken from it

- Like any form of artistic creation, restoration is a performing act that builds community capital and other transcendent values
- Restoration confronts idealized views of nature and of culture that are based on erasing that nature
- Human involvement with nature through restoration in no way besmirches the purity of nature
- Because restoration is a means for revealing nature's inner workings and exhibiting the values of benevolent landscape management rather than hiding them behind a conceptual mask or green barrier of "wilderness" preservation, it can be the ultimate form of environmental education
- By engaging local participation, restoration represents a positive, proactive and deliberate environmental act for concerned individuals tired of both the hands-off and non-active negativity associated with preservation or the widespread fatalism of just worrying about the state of the environment and the ensuing apathy that such a malaise can invoke
- Restoration, if approached free of the burden to copy or imitate what might have previously existed on a site, is a way to look forward into the future, not as has been suggested, limited by being yoked to the past
- Restoration, unlike preservation, blurs the distinction between culture and nature, freeing environmentalism of the polarizing (and false) human – nature dichotomy
- Restoration instills a respect for both the limited resilience and the inherent fragility of the land

B. Restoration as a Problematic or Detrimental Endeavor

- By giving humans opportunities to "play god," restoration removes mystery and limits nature to only that which humans can conceive
- Restoration projects may be technical accomplishments but are devoid of spirit or lasting significance
- Imitations of nature created by restorationists, like art created by performing chimpanzees, are impossible to treat seriously in that both are produced by agents who are in ignorance of what they do
- Though restoration projects tip their hats towards history in a pandering way, they are in fact new and condescending inventions devoid of authenticity; i.e. nothing is ever really restored
- Restorations are unnatural fakes and forgeries, and dangerous in that they offer deception as a placebo for the reality of environmental damage
- The originality of nature, its embodied history—the causal and evolutionary genesis of its accumulated ecological processes and landscapes—is critical in appraisal of its value, something absent in historically counterfeit restorations
- Believing in restoration is masquerading guilt behind a false and dangerous optimism of techno-fix hubris; i.e. if it breaks, no worry, we can always fix it…no need to change our behavior
- Restored nature is a pleasing illusion, a false commodity created to fulfill human goals, to appease human morality, and to sustain human sovereignty over and dominance of the natural world
- To dominate nature by engaging in restoration is to ultimately destroy it, an act made all the more dangerous because it is presented behind a guise of benevolence
- Far from being a solution to the problem, restoration is part of the problem; it is a compromise that at best superficially improves appearances, nothing more
- Because time flows forward and it is thus impossible to restore the past, restitution rather than restoration is the best means of moving ahead into a new relationship of humility with nature

C. Restoration as Relationship: Community Ritual and Design

- Restoration questions and challenges previous environmental assumptions by offering new rituals as means for transcending antiquated tradition
- Deep restoration is based on rebirth and redemption and is a symbolic act of performance to bring human and non-human communities together

- Restoration is a way of channeling nature's wisdom through human actions and imagination towards an end beneficial to both
- Restoration is a return to the roots of ecology as a healing art before it became a mechanistic discipline
- Restoration promotes education about how we have lived on the land and, therefore, about our place in nature and who we are; in short it represents the ultimate form of ecological literacy that creates not only new natures but also new deeper understandings about nature and strengthened desires to protect it
- If we believe that humans are part of, rather than apart from, nature, all human-aided landscape restorations are by definition "natural" exercises in applied history
- Ritual empowers restoration by creating a middle landscape and community of nature and civilization rooted in the past but designed toward the future
- Restoration, as a helpful metaphor for appropriate interventions in natural processes that create social capital, is an honest acknowledgment of our role as designers of both sociological and biological landscapes
- Because ecosystems are dynamic with no real temporal beginning, restoration can never return any natural system to any arbitrary "correct" past state; thus restoration works toward an idealized or even idolized version of imagined nature
- Robust restoration is a means of embedding historical legacy, both natural and cultural, within practical design
- Restoration as a design practice combines the processes of creating culture on a natural landscape as well as creating nature on a cultural landscape
- Mindful restoration balances historical fidelity and ecological integrity through social engagement with a focal practice
- A restored landscape is a re-storied landscape with a created sense of place based on memory
- Successful ecocultural restoration is accomplished only when health is returned to both the ecosystem and the human community
- Restoration design is a particular form of landscape architecture that integrates both art and science for the mutual benefits of both nature and culture

Sources

A & B are derived from Baldwin, A., J. De Luce and C. Fletsch (1994) *Beyond Preservation: Restoring and Inventing Landscapes.* Univ. Minn. Press; Gobster, P.H. and R.B. Hull (2000) *Restoring Nature: Perspectives form the Social Sciences and Humanities.* Island Press; and Throop, W. (2000) *Environmental Restoration: Ethics, Theory, and Practice.* Humanity Books.

C is derived from Jordan, W.R. (2003) *The Sunflower Forest: Ecological Restoration and the New Communion with Nature.* Univ. Calif. Press; and Higgs, E. (2003) *Nature by Design: People, Natural Process, and Ecological Restoration.* MIT Press.

Notes and References

Prologue: France

Cook, S. 2002. *The Great Swamp of Arlington, Belmont, and Cambridge: An historic perspective of its development* 1630-2001. Private Publicatoin

France, R.L. (Ed.) 2002. *Handbook of water sensitive planning and design.* CRC Press.

France, R.L. (Ed.) 2005. *Facilitating watershed management: Fostering awareness and stewardship.* Rowman & Littlefield.

France, R.L. 2005. Foreword by Series Editor: Blurring the border between nature and culture. *In* Vince, S.W., M.L. Duryea, E.A. Marie and L.A. Hermansen. (Eds.) *Forests at the wildland-urban interface: Conservation and management.* CRC Press.

France, R.L. 2006. *Introduction to watershed development: Understanding and managing the impacts of sprawl.* Rowman & Littlefield.

Metropolitan District Commission. 2003. *Alewife Reservation & Alewife Brook master plan.* MA Govt.

Pinkham, R. 2000. *Daylighting: New life for buried streams.* Rocky Mount. Inst.

Vilesisis, A. 1997. *Discovering the unknown landscape: A history of America's wetlands.* Island Press.

Introduction: France

Baldwin, A.D., J. De Luce and C. Fletsch. (Eds.) 1994. *Beyond preservation: Restoring and inventing landscapes.* Uni. Minn. Press.

Chard, P.S. 1999. *The healing earth: Nature's medicine for the troubled soul.* North Woods Press.

Clinebell, H. 1996. *Ecotherapy: Healing ourselves, healing the Earth.* Fortess Press.

Cohen, M.J. 1997. *Reconnecting with Nature: Finding wellness through restoring your bond with the Earth.* Ecopress.

Colegate, I. 2002. *A pelican in the wilderness: hermits and solitaries.* Counterpoint Press.

Cronon, W. (Ed.) 1996. *Uncommon ground: Rethinking the human place in nature.* W.W. Norton.

Cunningham, S. 2002. *The restoration economy: The greatest new growth frontier.* Berett-Koehler Publ.

Evernden, N. 1992. *The social creation of nature.* John Hopkins Univ. Press.

France, R.L. 1992. Gaian integrity: A clarion precept for global preservation. *The Trumpeter J. of Ecosophy* 9: 159-164.

France, R.L. 2001. (Stormwater) leaving Las Vegas. *Land. Arch. Mag.* 8: 38-42.

France, R. L. 2003. *Deep immersion: The experience of water.* Green Frigate Books.

France, R.L. 2005. (Ed.) *Facilitating watershed management: Fostering awareness and stewardship.* Rowman & Littlefield Publ.

France, R.L. 2006a. *Introduction to watershed development: Understanding and managing the impacts of sprawl.* Rowman & Littlefield Publ.

France, R. L. 2009. *Clark County Wetlands Park: Community-based restoration and open-space planning.* Island Press.

Gobster, P.H. and R.B. Hull. (Eds.) 2000. *Restoring nature: Perspectives from the social sciences and humanities.* Island Press.

Grossman, W. 1998. *To be healed by the Earth.* Seven Stories Press.

Higgs, E. 2003. *Nature by design: People, natural process, and ecological restoration.* MIT Press.

Hobbs, R.J., P.R. Ehrlich and D.A. Saunders. 1993. *Repairing a damaged world: An outline for ecological restoration.* Hyperion Books.

Jordan, W.R. 2003. *The sunflower forest: Ecological restoration and the new communion with nature.* Univ. Calif. Press.

Karr, J.R. and E.W. Chu. 1999. *Restoring life in running waters: Better biological monitoring.* Island Press.

Leopold, A. 1949. *A Sand County almanac with essays on conservation from Round River.* Ballatine Books.

Lovelock, J. 1991. *Healing Gaia: Practical medicine for the planet.* Harmony Books.

McHarg, I. L. 1999. *To heal the earth: Selected writings of I.L. McHarg.* Island Press.

McKibben, B. 1990. *The end of nature.* Anchor.

Marcus, C.C. 1999. *Healing gardens: Therapeutic benefits and design recommendations.* Wiley.

Merchant, C. 1989. *The death of nature.* HarperCollins.

Orr, D. 1992. *Ecological literacy: Education and the transition to a postmodern world.* SUNY Press.

Regier, H.A. and R.L. France. 1998. Perspectives on the meaning of ecosystem integrity in 1975. *In* Edwards, C.J. and H.A. Regier (Eds.) *An ecosystem approach to the integrity of the Great Lakes in turbulent times.* Great Lakes Fish. Comm.

Rozak, T., M.E. Gomes and A.D. Kanner. 1995. *Ecopsychology: Restoring the Earth, healing the mind.* Sierra Clubs Books.

Spelman, E.V. 2002. *Repair: The impulse to restore in a fragile world.* Beacon Press.

Steedman, R. J. 2005. Buzzwords and benchmarks: Ecosystem health as a management goal. *In* France, R.L., (Ed.) *Facilitating watershed management: Fostering awareness and stewardship.* Rowman & Littlefield Publ.

Takei, J. and M.P. Keane. 2002. *Sakuteiki: Visions of the Japanese garden, a modern translation.* Tuttle Publ.

Throop, W. (Ed.) 2000. *Environmental restoration: Ethics, theory, and practice.* Humanity Books.

Tuan, Y-F. 1977. *Space and place: The perspective of experience.* Univ. Minn. Press.

Tyson, M.M. 1998. *The healing landscape: Therapeutic outdoor environments.* McGraw-Hill.

United States Environmental Protection Agency. 1975. *The integrity of water.* US EPA.

Westra, L. and J. Lemons. 1995. *Perspectives on ecological integrity.* Kluwer Academic Publ.

Wordsworth, J., M.C. Jaye and R. Woof. 1987. *William Wordsworth and the age of English Romanticism.* Rutgers Univ. Press.

Chapter 1: Jordan and Turner

1. Anne W. Spirn, Constructing nature: The legacy of Frederick Law Olmsted *In* Cronon, W. *Uncommon Ground: Rethinking the Human Place in Nature* (New York: W.W. Norton, 1996) pages 91-113.

2. William R. Jordan III and Marcus Hall, "Ecological Restoration," *In* Shepard Krech , John R. McNeil and Carolyn Merchant, editors, *Encyclopedia of World Environmental History* (Great Barrington, MA: Berkshire Publishing Co., 2003)

3. Eric Katz. *Nature as Subject: Human Obligation and Natural Community* (Lanham, Maryland: Rowman& Littlefield, 1997), especially the essay "The Big Lie: Human Restoration of Nature," pp. 93-107, and the three essays that follow. The classic statement of the critique of restoration is Robert Elliot's, "Faking Nature," Inquiry 25: 81-93, 1982.

4. For an overview of Frederick Turner's ideas about value, see his *Beauty: The Value of Values* (Charlottesville: University of Virginia Press, 1991); and *The Culture of Hope: A New Birth of the Classical Spirit* (New York: The Free Press, 1995).

5. Joseph Campbell, *The Masks of God: Primitive Mythology* (New York: Penguin, 1959) page 181.

6. Interestingly, the Japanese do better. They have a word—*saitaien*—meaning "ecology park," which they use to refer to at least some restoration projects. See Toshihiko Nakamura, "Saitaien: Creating a collection of restored ecosystems in Chiba, Japan, *Restoration & Management Notes,* Summer, 1993 11(1) 25-30

7. In his dialogue poem "West-Running Brook"

8. Aldo Leopold, "Round River" in *A Sand County Almanac, With Essays from Round River* (New York: Ballantine, 1966), page 197

9. Philippe Descola and Gísli Palsson, editors, *Nature and Society: Anthropological Perspectives* (London: Routledge, 1996), especially the Introduction, pp 1-22, and Signe Howell's chapter, "Nature in Culture or Culture in Nature: Chewong Ideas of 'Humans" and other Species," pages 127-44.

10. Michael Shellenberger and Ted Nordhaus, The Death of Environmentalism:Global Warming Politics in a Post-Environmental World. This article, which has provoked considerable discussion in recent months, appeared on-line in the fall of 2004, and may be found at http://www.thebreakthrough.org/images/Death_of_Environmentalism.pdf.

11. J. Baird Callicott, "African biocommunitarianism and Australian dreamtime," Chapter 8, pages 156-172 IN *Earth's Insights: A Multicultural Survey of Ecological Ethics from the Mediterranean Basis to the Australian Outback.* (Berkeley: University of California Press, 1994).

12. Aldo Leopold, "The land ethic," pages 237-263, in *A Sand County Almanac, With Essays from Round River* (New York: Ballantine, 1966)

13. For a recent discussion of initiation and other coming-of-age traditions, see Ronald L. Grimes, *Deeply into the Bone: Reinventing Rites of Passage,* (Berkeley: University of California Press, 2000), especially Chapter 2, "Coming of Age, Joining up."

14. Konrad Lorenz, *On Aggression,* (New York: Bantam, 1967) pp. 63-64.

15. Frederick Turner, Bloody Columbus: Restoration and the transvaluation of shame into beauty. *Restoration & Management Notes,* 10(1) 70-74. The passage referred to is on page 73.

16. For an introduction to Turner's ideas about ritual see Victor Turner, *The Ritual Process: Structure and Anti-Structure,* (Ithaca, NY: Cornell University press, 1977). See especially Chapter 3, "Liminality and Communitas," and Chapter 4, "Communitas: Model and Process." See also Turner's general discussions of ritual in his *The Drums of Affliction: A Study of Religious Processes among the Ndembu of Zambia* (Oxford: Clarendon Press, 1968).

17. Catherine Pickstock, *After Writing: The Liturgical Consummation of Philosophy* (Oxford: Blackwell, 1998)

18. Aldo Leopold A Sand County Almanac: With Essays on Conservation from Round River, (New York: Ballantine Books,1978. Original edition: Oxford University Press, 1966. The passage referred to is on page 262.

19. Stephen R. Kellert and Edward O. Wilson, editors, *The Biophilia Hypothesis* (Covello, CA: Island Press, 1993)

20. Victor Turner, *The Forest of Symbols: Aspects of Ndembu Ritual* (Ithaca: Cornell University Press, 1967), page 20.

Chapter 2: Dannenhauer

Note: Eleven accompanying images are available from the author.

Allen, Timothy F. H. 1998. Community ecology: The issue at the center, in Dodson, Stanley I., T.F.H. Allen, S.R. Carpenter, A.R. Ives, R.L. Jeanne, J.F. Kitchell, N.E. Langston, M.G. Turner. 1998. Ecology. Oxford University Press.Berkes,

Fikret, C. Folke, J. Colding. 1998. Linking social and ecological systems: Management practices and social mechanisms for building resilience. Cambridge University Press.

Caplan, Ralph. 2005. By design: Why there are no locks on the bathroom doors in the Hotel Louis XIV and other object lessons, second edition. Fairchild Publications.

Chelsea Creek Restoration Partnership (CCRP). 2005. Chelsea Creek Restoration Partnership overview. http://www.noahcdc.org/chelsea_creek_overview.html

Colfax, Greg. 1994. Personal interview.

Comediants. 1988. Comediants: 15 Anos. Centro de Documentation Teatral.

Doran, Bob. 2003. Political theater: Petrolia couple are challenging perceptions. North Coast Journal: October 23, 2003, online version.

Foster, David R. 1999. Thoreau's country: Journey through a transformed landscape. Harvard University Press.

Gilbert, O.L. 1991. The ecology of urban habitats. Chapman and Hall.

Gobster, Paul H., R.B. Hull. 2000. Restoring nature: Perspectives from the social sciences and humanities. Island Press.

Gonick, Gloria Granz. 2002. Matsuri: Japanese Festival Arts. UCLA Fowler Museum of Cultural History.

Green, Susan. 1985. Bread and Puppet: Stories of struggle & faith from Central America. Green Valley Film and Art.

Higgs, Eric. 2003. Nature by design: People, natural process, and ecological restoration. MIT Press.

House, Freeman. 1999. Totem salmon: Life lessons from another species. Beacon Press.

Huet, Michel. 1978. The dance, art and ritual of Africa. Pantheon Books.

Krause, Bernie. 2002. Wild soundscapes: Discovering the voice of the natural world. Wilderness Press.

Lommel, Andreas. 1972. Masks: Their meaning and function. McGraw-Hill.

Luoma, Jon R. 1999. The hidden forest: The biography of an ecosystem. Henry Holt and Company.

Nabhan, Gary Paul. 1997. Cultures of habitat: On nature, culture, and story. Counterpoint.

Nimis, Pier Luigi, C. Scheidegger, P.A. Wolseley. 2002. Monitoring with lichens — Monitoring lichens. Kluwer Academic Publishers.

Nunley, John W., J. Bettelheim. 1988. Caribbean Festival Arts. The Saint Louis Art Museum and University of Washington Press.

Roberts, Allen F.1995. Animals in African art: From the familiar to the marvelous. The Museum for African Art.

Society for Ecological Restoration Science & Policy Working Group (SER). 2002. The SER Primer on Ecological Restoration. www.ser.org/.

Shiva, Vandana. 1997. Biopiracy: The plunder of nature and knowledge. South End Press.

Shiva, Vandana. 1993. Monocultures of the mind: Perspectives on biodiversity and biotechnology. Zed Books.

Teele, Rebecca. 1984. Mime Journal, 1984: No/Kyogen masks and performance. Pomona College Theater Department.

Tufte, Edward R. 1998. Visual explanations: Images and quantities, evidence and narrative. Graphics Press.

Upton, Dell. 1998. Architecture in the United States. Oxford University Press.

Chapter 3: Conn and Conn

Abram, D. (1996) *The spell of the sensuous: Perception and language in the more-than-human world.* New York: Pantheon.

Arheim, R. (1961) "The Gestalt theory of expression," Henle, M.A., Ed. *Documents in Gestalt Psychology.* Berkeley: U.C. Press.

Armstrong, J. (1995) "Keepers of the earth," in T. R. Roszak, M.E. Gomes, A. D. Kanner, Eds., *Ecopsychology: Restoring the earth, healing the mind.* San Francisco: Sierra Club, pp. 316-324.

Bateson, G. (1972) *Steps to an ecology of mind.* New York: Ballantine.

Berry, T. (1999) *The great work.* New York: Bell Tower.

Bly, R. (1980) *News of the universe: Poems of twofold consciousness.* San Francisco: Sierra Club.

Bortoft, H. (1996) *The wholeness of nature: Goethe's way toward a science of conscious participation in nature.* New York: Lindisfarne.

Brown, M.Y., Ed. (1994) *Lighting a Candle: Quotations on the Spiritual Life.* Center City, Minn.: Hazeldon.

Capra, F. (1988) *Uncommon wisdom: Conversations with remarkable people.* New York: Simon & Shuster.

Conn, L.K. (2002) "Opening to the other," unpublished manuscript.

Conn, S.A. (1996) *Ecopsychology.* (Video) Palo Alto, CA: Foundation for Global Community.

Conn, S.A. and L.K. Conn (1998) "Ecopsychology and psychotherapy within the larger context," *Gestalt Review,* 3(2): 119-125.

Davis, J. (2000) "A stab at a definition of ecopsychology," unpublished notes from a gathering of ecopsychologists in Boulder, CO., August.

DeQuincey, C. (2002) "Stories matter, matter stories," *IONS: Noetic sciences review.* June-August, No. 60, 8-13, 44-45.

Fisher, A. (2002) *Radical ecopsychology: Psychology in the service of life.* Albany: State University of New York.

Flemons, D. G. (1991) *Completing distinctions.* Boston: Shambhala.

Heidegger, M. (1968) Trans. by F.D. Wieck & J. Gray. *What is called thinking?* New York: Harper & Row.

Heidegger, M. (1971) Trans. by P. D. Hertz. *On the Way to Language.* New York: Harper & Row.

Heidegger, M. (1972) Trans. by J. Stambaugh. *On time and being.* New York: Harper & Row.

Huxley, A. (1954) *The doors of perception.* New York: Harper & Row.

Jensen, D. (2002) *Interviews: Listening to the land.* New York: Context.

Kidner, D. W.(2001) *Nature and psyche: Radical environmentalism and the politics of subjectivity.* New York: SUNY

Kimbrell, A. (2002) *Cold evil: Technology and modern ethics.* Barrington, MA: Schumacher Society Lectures.

Koestler, A. (1978) *Janus: A summing up.* New York: Random House

Lane, R. (1998) "A Council of all Beings: The site speaks," in Macy, J. & Brown, M.Y. *Coming back to life: Practices to reconnect our lives, our world.* Stony Creek, CT: New Society, 207-210.

Latner, J. (1992) "The theory of gestalt therapy," in Nevis, E. C., Ed. *Gestalt therapy:Perspectives and applications.* New York: Gardner, 13-56.

Macy, J. & Brown, M.Y. (1998) *Coming back to life: Practices to reconnect our lives, our world.* Stony Creek, CT: New Society.

Narby, J. (1998) *The cosmic serpent: DNA and the origins of knowledge.* New York: Jeremy Tarcher.

Nhat Hanh, Thich. (1987) *Interbeing.* Berkeley, CA: Parallax.

Odum, E. (1971) *The fundamentals of ecology.* Philadelphia.: W. B. Saunders.

Ornstein, R. E. (1983) *The mind field.* London: Octagon Press.

Roszak, T. (1992). *The voice of the earth.* New York: Simon and Schuster.

Seed, J; Macy, J; Fleming,P; Naess, A. (1988) *Thinking like a mountain: Towards a council of all beings.* Philadelphia, Pa.: New Society.

Shepard, P. (1973). *The tender carnivore and the sacred game.* New York: Charles Scribner's Sons.

Shepard, P. (1982) *Nature and madness.* San Francisco: Sierra Club.

Shepard, P. (1995) "Virtually hunting reality in a forest of simulacra," Soule, M.E. & Lease, G., Eds. (1995) *Reinventing nature?: Responses to postmodern deconstruction.* Washington, D.C.: Island Press, pp 17-30.

Snyder, G. (1992) *No nature.* New York: Pantheon.

Soule, M.E. & Lease, G., Eds. (1995) *Reinventing nature?: Responses to postmodern deconstruction.* Washington, D.C.: Island Press.

Swimme, B. and Berry, T.(1992) *The universe story.* San Francisco: Harper.

Taussig, M. (1980) *The devil and commodity fetishism in South America.* University of N. Carolina Press.

Tolle, E. (1999) *The power of now.* Novato, CA: New World Library.

Van der Ryn, S. and Stuart Cowan. (1995) *Ecological design.* Washington, D.C.: Island Press.

Watts, A. (1972) *The Book: On the taboo against knowing who you are.* New York: Vintage.

Watts, A. (1991) *Nature, man and woman.* New York: Vintage.

Whitehead, A.N. (1955). *Adventures of Ideas.* New York: New American Library.

Whitehead, A.N. (1956). *Science and the modern world.* New York: New American Library.

Wilber, K. (1981) *Up from Eden: A Transpersonal view of human evolution.* Boulder, CO: Shambala.

Williams, T. T. (2002) *RED: Passion and patience in the desert.* New York: Vintage.

Winter, D. D.N. (1996) *Ecological psychology: Healing the split between planet and self.* New York: Harper Collins

Zimmerman, M. E. (1981) *Eclipse of the self: The Development of Heidegger's concept of authenticity.* Athens, OH: Ohio University Press.

Chapter 4: Kidner

(1) Alfred N. Whitehead, *The Concept of Nature* (Cambridge: Cambridge Univ. Press, 1920), p. 73

Christopher Lasch, *The culture of Narcissism* (New York: Norton, 1979).

(2) Personal Communication.

(3) John Rodman, "The liberation of nature?". Inquiry 20 (1977), p. 104.

(4) Rodman, "The liberation of nature?", p. 104.

(5) Rodman, "The liberation of nature?", p. 104.

(6) Gary Nabhan, "Culture parallax in viewing North American habitats". In: Micheal E. Soule and Gary Lease (eds.), *Reinventing Nature?: Responses to Postmodern Deconstruction.* (Washing DC: Ilsan Press, 1995), p. 98

(7) See, for example, *my Nature and Psyche: Radical Environmentalism and the Politics of Subjectivity* (Alban: Sate University of New York Press, 2001). See also: Tim Luke, "Green consumerism: Ecology and the ruse of recycling", in Jane Bennett and William Chaloupka (eds.), *In the Nature of Things: Language, Politics, and the Environment* (Minneapolis: University of Minnesota Press, 1993); Jennifer Price, "Looking for nature at the mall: A field guide to the nature company", in William Cronon (ed.), *Uncommon Ground: Rethinking the Human place in Nature* (new York: Norton, 1995).

(8) Roger Brooke, *Jung and Phenomenology* (London: Routhledge, 1991), p. 109.

(9) Chapter 6 by Mills considers the importance of historical reinhabitation of a landscape.

(10) See also Chapger 7 by Spelman, who discusses the dangers of such conceptual 'time travel'.

(11) Robin Ridington, Trail to Heaven: *Knowledge and Narrative in a Northern Native Community* (Vancouver: Douglas and McIntyre, 1988), p. 70, 72.

(12) Daniel Nettle and Suzanne Romaine, *Vanishing Voices: The Extinction of the World's Languages* (Oxford: Oxford University Press, 2000), p. 74

(13) The Jungian concept of archetype raises analogous questions of the relation between past and present in the psychological realm.

(14) Blade Runner, directed by Ridley Scott.

(15) Christopher Lasch, *The Culture of Narcissism* (New York: Norton, 1979), pp. 5.

(16) See Clifford Geertz, The Interpretation of Cultures (New York: Basic Books, 1973), for a critique of the 'stratigraphic' conception of the psyche.

(17) Sigmund Freud, *Civilisation and its Discontents.* (London: Hogarth, 1952).

(18) Rodman, The liberation of nature?:, p. 105.

(19) Jung and Phenomenology, p. 60-61.

(20) R. Bruce Hull and David P. Robertson, "The Language of Nature Matters: We Need a More Public Ecology". In Paul H. Gobster and R. Bruce Hull, *Restoring Nature: Perspective from the Social Sciences and Humanities* (Washington.Dc: Island Press, 2000), p. 106.

(21) Hull and Robertson, "The Language of Nature Matters", p. 102.

(22) Hull and Robertson, "The Language of Nature Matters", p. 104.

(23) William M. Denevan, "the pristine myth: The landscape of the Americas in 1492". Annals of the Association of American Geographers 82(3), 1992, 369-385.

(24) Hull and Robertson, "The Language of Nature Matters", p. 102.

(25) Hull and Robertson, "The Language of Nature Matters", p. 102.

(26) Hull and Robertson, "The Language of Nature Matters:, p. 104.

(27) Michael Fischer, Review of Christopher Lasch, "The Culture of Narcissism". Salmagundi, Fall 1979, 166-173.

(28) Peter Mason, *Deconstructing America: Representations of the Other* (London: Routledge, 1990), p. 15.

(29) Wade Sikorski, "Building wilderness"; in Jane Bennett and William Chaloupka (eds.), *In the Nature of Things: Language, Politics, and the Environment* (Minneapolis: University of Minnesota Press, 1993), p. 29.

(30) William Cronon, "The trouble with wilderness; or, Getting back to the wrong nature". In William Cronon (ed.), *Uncommon Ground,* p. 69.

(31) See my "Fabricating Nature: A Critique of the Social Construction of Nature". *Environmental Ethics,* 22 No. 4, Winter 2000, 339 – 357.

(32) Hull and Robertson, "The Language of Nature Matters, p. 106.

(33) T. G. H. Strehlow, *Songs of Central Australia* (Sydney: Angus and Roberston, 1971), p. 9.

(34) Tim Ingold, "The optimal forager and economic man", in Philippe Descola and Gisli Palsson (eds.), *Nature and Society: Anthropological Perspectives* (London: Routledge, 1996), p. 26.

(35) See my "Nature and Human Intelligence", *Human Ecology Review* 6(2), Winter 1999, 10-22

(36) Toni Morrison, "The Site of Memory", in William Zinsser (ed.), *Inventing the Truth: The Art of Craft of Memoir* (New York: Houghton Mifflin, 1987), p. 119.

(37) Morrison, "The Site of Memory", pp. 111-2.

(38) Morrison, "The Site of Memory", pp. 113.

(39) By anology, ponder your reaction if it was suggested that your spouse/lover should be replaced by a genetically identical clone. And why does this idea not appeal …?

(40) Jack Turner, *The Abstract Wild* (Tuscon: University of Arizona Press, 1996), p. 15.

(41) Aldo Leopold, "Wilderness". Quoted by J. Baird Callicot, "the wilderness idea revisited: The sustainable development alternative". *The Environmental Professional* 13 (1991), p. 238.

(42) See, for example, Eric Katz, "The big lie: Human restoration of nature". In William Throop (ed.), *Environmental Restoration; Ethics, Theory, and Practice* (Amherst: Humanity Books, 2000).

(43) Tim Weiskel, address to the 'Brown Fields and Gray Waters' conference, Harvard University, November 8th – 11th 2001.

(44) See, for example, the National Oceanographic and Atmospheric Technical Memorandum NORSORCA 125: "Oil spill impacts and the biological basis for response guidance: An applied synthesis of research on three subarctic intentional communities" (Seattle, Washington: March 1998); and the National Oceanographic and Atmospheric Technical Memorandum NORSORCA 114: "Integrating physical and biological studies of recovery from the Exxon Valdez oil spill" (Seattle, Washington: September 1997).

(45) Andrew Light and Eric Higgs, "The Politics of Ecological Restoration". Environmental Ethics 18 (1996), 230.

(46) See Greg Philo and David Miller, *Market Killing: What the Free Market Does and What Social Scientists Can Do About It* (Harlow, Essex: Pearson Education, 2001), for a review of this topic.

(47) See my "Culture and the Unconscious in Environmental Theory", Environmental Ethics, 20 (1998), 61 – 80.

(48) Hull and Robertson, "The Language of Nature Matters", p. 101.

(49) See William M. Schaffer, "Stretching and folding in lynx returns: Evidence for a strange attractor in nature?". *American Naturalist* 124 (1984), No. 6, 798 – 820. I refer the reader to James Gleick, Chaos (London: Heinemann, 1988), for an accessible introduction to chaos theory.

(50) Hull and Robertson, The Language of Nature Matters", p. 105.

(51) Hull and Robertson, The Language of Nature Matters", p. 104.

(52) Hull and Robertson, The Language of Nature Matters", p. 114.

(53) Hull and Robertson, The Language of Nature Matters:, p. 115.

(54) Robert L. Ryan, "A People-Centered Approach to Designing and Managing Restoration Projects: Insights from Understanding Attachment to Urban Natural Areas". In Paul H. Gobster and R. Bruce Hull, *Restoring Nature: Prespectives from the Social Sciences and Humanities* (Washington, DC: Island Press, 2000), p. 102.

(55) Ryan, "A People-Centered Approach to Designing and Managing Restoration Projects", p. 213.

(56) See, for example, R. A. Spitz, "Anaclitic Depression", *Psychoanalytic Study of the Child* 2 (1946): 313 – 342; John Bowlby, *Separation: Anger and Anxiety* (London: Tavistock, 1973).

(57) See, for example, Cisco Lassiter, "Relocation and Illness: The Plight of the Navaho", in David Michael Levin (ed.), *Pathologies of the Modern Self: Postmodern Studies on Narcissism, Schizophrenia, and Depression* (New York: New York University Press, 1987).

(58) On the distancing effect of a technological perspective, see Robert Romanyshyn, *Technology as Symptom and Dream* (London: Routledge, 1989). I am not, of course, suggesting that we should abandon thinking and replace it by feeling; only that these two modes of being should both exist within a healthy person.

Chapter 5: Light

Brennan, Andrew 1998. "Poverty, Puritanism and Environmental Conflict," *Environmental Ethics* 7: 305-331.

Elliot, Robert 1997. Faking Nature. London: Routledge.

Elliot, Robert 1982. Faking Nature, " *Inquiry* 25: 81-93.

Gunn, Alastair 1991. "The Restoration of Species and Natural Environments," Environmental Ethics 13: 291-309

Katz, Eric 2002. "Understanding Moral Limits in the duality of Artifacts and Nature: A Reply to Critics," *Ethics and the Environment* 7: 138-146.

Katz, Eric 1997. *Nature as Subject: Human Obligation and Natural Community.* Lanham, MD: Rowman & Littlefield Publishers.

Katz, Eric 1996. "The Problem of Ecological Restoration," *Environmental Ethics* 18: 222-224.

Light, Andrew 2003a. "Urban Ecological Citizenship" *Journal of Social Philosophy* 34.

Light, Andrew 2003b. "'Faking Nature' Revisited." In D. Michelfelder and B. Wilcox (eds.) *The Beauty Around Us: Environmental Aesthetics in the Scenic Landscape and Beyond.* Albany, NY: SUNY Press.

Light, Andrew 2002a. "Contemporary Environmental Ethics: From the Metaethics to Public Philosphy," *Metaphilosophy* 33: 426-449.

Light, Andrew 2002b. "Restoring Ecological Citizenship," In B. Minteer and B. P. Taylor (eds.) *Democracy and the Claims of Nature.* Lanham, MD: Rowman & Littlefield, 153-172.

Light, Andrew 2001. "The Urban Blind Spot in Environmental Ethics," *Environmental Politics* 10: 7-35.

Light, Andrew 2000a. "Restoration, the Value of Participation, and the Risks of Professionalization." In P. Gobster and B. Hull (eds.) *Restoring Nature: Perspectives from the Social Sciences and Humanities.* Washington, DC: Island Press, 49-70.

Miles, I., et. Al.. 2000. "Psychological Benefits of Volunteering for Restoration Projects." *Ecological Restoration* 18, 218-227.

Rolston, Holmes III. 1994. Conserving Natural Value. New York: Columbia University Press.

Scheffler, Samuel 1997. "Relationships and Responsibilities," *Philosophy and Public Affairs* 26: 189-209.

Scherer, Donald 1995, "Evolution, Human Living, and the Practice of Ecological Restoration," *Environmental Ethics* 17: 359-379.

Stevens, William K. 1995. Miracle Under the Oaks. New York: Pocket Books.

Throop, William (1997). "The Rationale for Environmental Restoration." In. R. Gottlieb (ed.) *The Ecological Community.* London: Routledge, 39-55.

Wilson, Edward O. 1992. *The Diversity of Life.* New York: W. W. Norton and Company

Chapter 6: Mills.

(1) Stephanie Mills, *Whatever Happened to Ecology?* (San Francisco: Sierra Club Books, 1989).

(2) Peter Berg and Raymond F. Dasmann, *"Reinhabiting California"* in Reinhabiting A Separate Country, Peter Berg, Ed.

(San Francisco: Planet Drum Books, 1978), pp. 217-218.

(3) Stephanie Mills, *In Service of the Wild: Restoring and Reinhabiting Damaged Land.* (Boston: Beacon Press, 1995).

(4) Aldo Leopold, *A Sand County Almanac with Essays On Conservation From Round River.* (New York: Sierra Club Ballantine Books, 1966. Published by arrangement with the Oxford University Press), p.197.

(5) Aldo Leopold, *A Sand County Almanac,* p. 158.

(6) Joseph A. Tainter, *The Collapse of Complex Societies.* Cambridge:Cambridge University Press, New Studies in Archaeology, 1988).

(7) Deanne Urmy, flaps copy on *In Service of the Wild* by Stephanie Mills.

(8) Aldo Leopold, *A Sand County Almanac,* p. 262.

(9) Carolyn Raffensperger, "Precaution and Security: The Labyrinthine Challenge" in *Whole Earth* 3109, Fall 2002, p.34.

(10) Stephanie Mills, Ed. *Turning Away From Technology: A New Vision for the Twenty-first Century.* (San Francisco; Sierra Club books, 1997).

(11) Stephanie Mills, "Making Amends to the Myriad Creatures". (Eleventh Annual E.F.Schumacher Lectures. Great Barrington, MA: E.F. Schumacher Society, 1991).

Chapter 7: Spelman

(1) Some passages of this essay are taken from my Repair: *The Impulse to Restore in a Fragile World.* Boston: Beacon Press, 2002.

(2) The vastness of our reparative repertoire is readily and richly attested to by sources ranging from *Reader's Digest* stories about legendary handymen, to books on the ethics of environmental restoration; from Dave Barry's send-ups of men's delusions about their superior repairing skills, to legal treatises weighing monetary reparations against the work of truth and reconciliation commissions; from *The you don't need a man to fix it book: The Woman's Guide to Confident Home Repair,* to *Tikkun,* the journal emblazoned with the Hebrew phrase *tikkun olam,* to repair the world. Newspapers and magazines provide a steady stream of reports on the vast variety of projects and problems awaiting *H. reparans:* former President Bill Clinton has "a reputation to salvage," though he and former Vice President Al Gore are said to have "patched up their tattered relationship." Citizens in Cincinnati, Ohio are working on "repairing civic morale" in response to heightened racial tensions. A team of conservators at the Stedelijk Museum in Amsterdam has successfully met "one of the biggest challenges of their profession: how to repair, seamlessly, a large-format, basically monochromatic canvas" (Barnet Newman's *Cathedra*). *Consumer Reports* regularly offers advice on whether to "Fix it or nix it," "Fix it or sell it." The wide range of responses to the horrible wounds inflicted on September 11, 2001, bear solemn witness to the sheer variety of *H. reparans'* capacities: the twin towers can neither be repaired nor restored, but as the President of the Historic Districts Council of New York City saw it, whatever is done at the site "must reweave the damaged threads of fabric that terrorism sought to tear apart, and create a sense of place that fills the void and honors the losses of Sept. 11." In one issue alone of the *New York Times Magazine,* there were stories devoted to the tasks of "mending a psyche" and of figuring out "how to put the family back together." An op-ed essay by former Secretary of the Treasury Robert Rubin on September 30 described "A Post-Disaster Economy in Need of Repair." See Pat Walsh, "Mr. Rhoades's Neighborhood." *Reader's Digest,* January 1995, 62-66; Environmental *Restoration: Ethics, Theory, and Practice,* ed. William Throop (Amherst, NY: Humanity Books, 2000); "The Tool Man Cometh." *Ellsworth American,* June 27, 1997, 9-10; *Between Vengeance and Forgiveness: Facing History After Genocide and Mass Violence,* by Martha Minow (Beacon, 1998) and *Interracial Justice: Conflict & Reconciliation in Post-Civil Rights America,* by Eric K. Yamamoto (NYU Press, 1999); *The you don't need a man to fix it book,* by Jim Webb and Bart Houseman, with an introduction by Erma Bombeck (Garden City: Doubleday, 1973); Chris Black, "Where's Bill? Fiddling with his legacy – and his party." *The American Prospect,* February 25, 2002, 12; Francis X. Clines, "A City Tries to Turn Candor Into Consensus." *The New York Times,* September 9, 2001, 16; Carol Vogel, "Restored, but Still Blue." *The New York Times,* January 4, 2002, B40; *Consumer Reports,* cover titles August 2000, October 2001; Hal Broom, Letters, *The New York Times,* November 10, 2001; Susan Dominus, "Mending a Psyche," and Lisa Belkin, "Life Without Father." *The New York Times Magazine,* November 11, 2001, 69, 112; Robert E. Rubin, "A Post-Disaster Economy in Need of Repair." *The New York Times,* September 30, 2001, 13.

(3) Review of Cheryl Mendelson's Home Comforts: *The Art and Science of Keeping House* (Scribner, 2000) in *The New Yorker,* July 24, 2000, 75.

(4) *Manchester Guardian Weekly,* September 15, 1996, 12.

(5) Andrew Pollack, "Digital Film Restoration Raises Questions About Fixing Flaws," *The New York Times*, March 16, 1998, D1.

(6) Katha Pollitt, "What's Right About Divorce," *The New York Times*, June 27, 1997, A 29.

(7) Lebbeus Woods, *Radical Reconstruction* (New York: Princeton Architectural Press, 1997).

(8) Aldo Leopold, *A Sand County Almanac, with Other Essays on Conservation from Round River* (New York: Oxford University Press, 1966), 183.

(9) See for example Robert Harbison, *The Built, the Unbuilt and the Unbuildable: In Pursuit of Architectural Meaning* (Cambridge, MA: MIT Press, 1991), 121-130; Camilo José Vergara, *American Ruins* (New York: Monacelli Press, 1999).

(10) For a recent description of the use of architecture in the control of its inhabitants, see Richard Sennett, *Respect: The Formation of Character in an Age of Inequality* (London: Penguin Books, 2003), 10 ff.

(11) Neil Leach, *The Anaesthetics of Architecture* (Cambridge, MA: The MIT Press, 1999), 27-32.

(12) Yuriko Saito, "The Japanese Aesthetics of Imperfection and Insufficiency." *The Journal of Aesthetics and Art Criticism* 55:4 Fall 1997, 377-385.

(13) New Haven: Yale University Press, 1991.

(14) Langer, *Holocaust Testimonies*, p. 136 (quoting survivor Julia S.).

(15) Lawrence L. Langer, *Pre-Empting the Holocaust* (New Haven: Yale University Press, 1998), 108.

(16) Lisa Lowe and David Lloyd, "Introduction." *The Politics of Culture in the Shadow of Capital*, ed. Lowe and Lloyd (Durham: Duke University Press, 1997), 27. Lowe and Lloyd here are insisting that however thoroughly culture has been commodified, it still "constitutes a site in which the reproduction of contemporary capitalist social relations may be continually contested." (26). Many thanks to George Lipsitz for the reference.

(17) As noted earlier, "repair" and "restore" are not synonymous. While the restoration of a building, for example, to an earlier stage of its history involves repair, repairing a building so that it can carry out its basic functions does not involve restoration to an earlier structural or aesthetic state.

(18) James Marston Fitch, *Historic Preservation: Curatorial Management of the Built World* (Charlottesville and London: University Press of Virginia, 1990), 46.

(19) *Art and Technics* (Columbia University Press, 1952), 42.

(20) Jane Perlez, "Decay of a 20th Century Relic: What's the Future of Auschwitz?" *The New York Times*, January 5, 1994, A6.

(21) Many thanks to Robert France, Martha Minow, and Monique Roelofs for helpful criticisms of earlier drafts.

Chapter 8: Brown

Note: Eight accompanying images are available from the author.

(1) Helen Mayer Harrison and Newton Harrison, "On the Endangered Meadows of Europe, 1996," (Conference on Meadows, Del Mar, CA: Harrison Studio, 1996) 7.

(2) Barbara Matilsky, *Fragile Ecologies: Contemporary Artists' Interpretations and Solutions* (New York: Rizzoli in association with the Queens Museum of Art, 1992) 57.

(3) Patricia Johanson, "Presentation: Art and Survival by Patricia Johanson," *Waterworks: A Symposium on Art and Water* (Cambridge: Cambridge Arts Council at Harvard University, 5 Apr. 1997).

(4) Qtd. in Eleanor Munro, *Originals: American Women Artists* (New York: Simon and Schuster, 1982) 462.

(5) Walter R. Davis, letter to Patricia Johanson, 24 Jan. 1982.

(6) Johanson, *Art and Survival: Creative Solutions to Environmental Problems* (North Vancouver, B.C., Canada: Gallerie Publications, 1992) 32.

(7) Lucy Lippard, Patricia Johanson: *Fair Park Lagoon, Dallas and Color Gardens* (New York: Rosa Esman Gallery, 1993) 12.

(8) *Village Voice*, qtd. in *Art in the Land* 151.

(9) Qtd. by Robin Cembalest in "The Ecological Art Explosion," *Artnews* (Summer 1991): 100.

(10) Alan Sonfist, *Sculpting with the Environment: A Natural Dialogue* ed. Baile Oakes (New York: Van Nostrand Reinhold, 1995) 159.

(11) Lippard, *Overlay: Contemporary Art and the Art of Prehistory* (New York: Pantheon Books, 1983) 231.

(12) Helen Harrison, telephone interview, 1 July 1997.

(13) Helen and Newton Harrison, "Shifting Positions toward the Earth: Art and Environmental Awareness," *Leonardo* 26 (1993): 371.

(14) Qtd. by Jure Mikuz in "Art and Ecology Today," *Breathing Space for the Sava River: A Proposition in Conversation with Hartmut Ern and Martin Schneider-Jacoby* (European Nature Heritage Fund, 1990) n.p.

(15) Qtd. by Michael Auping in *Common Ground: Five Artists in the Florida Landscape* (Sarasota, FL: John and Mable Ringling Museum of Art, 1982) 33.

(16) Martin Schneider-Jacoby, letter to the author, 2 December 2002.

(17) Jody Pinto, *Sculpting* 25-28.

(18) Pinto, *Sculpting* 25.

(19) Pinto, telephone interview, 9 June 1997.

(20) Pinto, *Sculpting* 28.

(21) Pinto, telephone interview, 9 June 1997.

(22) John Beardsley, *Earthworks and Beyond* (New York: Abbeville Press, 1984) 10.

(23) Beardsley, "The Intersection of Sculpture and Landscape Architecture," audio recording (Providence: International Sculpture Conference, June 1996).

(24) Robin Cembalest, *"Lettuce in a Landfill,"* Nation (Summer 1991): 43.

(25) Lippard, *Overlay* 217.

(26) Qtd. in "Revival Field," *Greenkeeping* (May-June 1992): 35.

(27) Cembalest, *Nation* 43.

(28) Chaney,telephone interview, 2 August 2002.

(29) Johanson, interview.

(30) Davis, letter.

(31) Danny Hobson, letter to the author, 10 April 2003.

(32) Eleanor Heartney, "Ecopolitics/Ecopoetry: Helen and Newton Harrison's Environmental Talking Cure," *But Is It Art?: The Spirit of Art as Activism*, ed. Nina Felshin (Seattle: Bay Press, 1995) 143.

(33) Cembalest, "The Ecological Art Explosion," *Artnews* (Summer 1991): 101.

(34) Suzi Gablik, *Conversations before the End of Time: Dialogues on Art, Life and Spiritual Renewal* (London: Thames and Hudson, 1995) 102.

(35) Jack Burnham, *Great Western Salt Works: Essays on the Meaning of Post-Formalist Art* (New York: George Braziller, 1974); Irene Diamond and Gloria Feman Orenstein, eds. *Reweaving the World: The Emergence of Ecofeminism* (San Francisco: Sierra Club Books, 1990); Gablik, *Has Modernism Failed?* (London and New York: Thames and Hudson, 1984); Lippard, *Overlay.*

(36) Patricia Sanders, "Eco-Art: Strength in Diversity," *Art Journal* 51 (Summer 1992): 79.

(37) Gablik, *Conversations* 133-54.

(38) Lippard, "The Long View: Patricia Johanson's Projects, 1969-86," *Patricia Johanson: Drawings and Models for Environmental Projects*, 1969-1986, ed. Debra Bricker Balken (Pittsfield: Berkshire Museum, 1987) 11.

Chapter 9: Collins

(1) The argument for this was made by Anne Winston Spirin during a pre-conference discussion on Healing Nature at the Brown Fields and Gray Waters Conference held at the Harvard Graduate School of Design on November 9, 2001.

(2) The modernist goals of creative autonomy and free creativity resulted in a discipline without traditions, so heavily invested in discipline specific innovation that it was eventually entropic. See Gablik (Art After Modernism: 1984) and Arthur Danto (After the End of Art: 1995) for a good overview of this phenomenon, and the authors argument that art must reengage social and environmental issues. Gablik extends her argument, with a strong focus on the environment in (The Reenchantment of Art: 1989)

(3) Joseph Beuys, was a German artist internationally recognized for his art and his social activism. Beuys had a role in founding the German student party, and the German Green Party. One of his last works, 7000 Oaks for Kassel Germany, began with the dumping of 7000 basalt columns in front of the primary exhibition building at Documenta 7, in 1982. His intention was to pair the columns with Oak tress to be placed/planted throughout Kassel. The Work, an act of urban regeneration and nature/culture restoration took 5 years to complete. (http://www.diacenter.org/ltproj/7000/). The legacy of Beuys includes both practice and theory.. He is best known for his idea of social sculpture whereby we –all- (everyone is an artist) must take responsibility for shaping the world in which we live. The shift from Beuys theory to the theoretical and philosophical position of todays neo-Beuysian practitioners can be described in terms of a post authorship practice. Where Beuys, retained his role as primary author, the new practitioners share authorship with a focus on effecting change. The clearest example of this can be found in the Autstrian group Wochenklausur . (http://www.wochenklausur.at/projekte/menu_en.htm)

(4) Helen and Newton Harrison are best known in the U.S. for the work on the "Lagoon Cycle" A 12 year study of coastal lagoons. (http://www.communityarts.net/readingroom/archive/ca/raven-harrison.php) Their work focuses upon the co-evolution of biodiversity and cultural diversity. They have worked all over the world developing visionary plans for the restoration of major rivers systems such as the Sava River in former Yugoslavia, the North American Rainforest – the redwoods of the northwest, and their most recent project Peninsula Europe, a study of the uplands of Europe and their ecological and cultural value for the entire European Union. Still active internationally, the Harrison's have recently wrote about their own work, at (http://moncon.greenmuseum.org/papers/harrison1.html). Their work with ecosystems and creativity; discourse and policy; at a planning scale is held in high regard. Their ongoing work affects the ideas and practices of artists like Aviva Rahmani in Maine, Ala Plastica in Argentina, and David Haley in England.

Other important contbitutors include Hans Haacke, Alan Sonfist and Agnes Denes. Haacke explored plants, natural phenomena and the water quality of the Rhine. In New York, Alan Sonfist proposed the restoration of a native forest to parklands throughout Manhattan, resulting in the "Time Landscape" in SoHo. Agnes Denes grew wheat at Battery Park City beneath the shadows of the twin towers. Shifting a brownfield site from a wasteland, "Wheatfield" became a symbolic source of wheat and bread for a city that had long forgotten its relationship to agriculture. In 1974, Jack Burnham wrote an important book that featured many of these artists. Titled, "Great Western Salt Works" (1974: 15-24), it was notable because it developed an initial approach to systems aesthetics.

(5) Other Eco-art Dialgoue members contributing to this definition and guideline include Lynne Hull, of Colorado, one of the most consistent practitioners of "trans-species" art, or work with/for animals and wildlife. Aviva Rahmani of Vinalhaven, Maine has spent nine years restoring a tidal wetland on property she owns there. Susan Liebovitz Steinman of Oakland, California, has recently finished a large collaborative ecological restoration and planning project working with the U.S. Park Service. Artists Ann Rosenthal of Massachusetts, Jackie Brookner of New York (Steinman's collaborator on the Park Service Project.) and the curator Amy Lipton of New York have provided additional input and support for this effort and its realization. Others members that are active on the list include, include the artists Jeroen Van Westen of Holland, Shai Zakai of Israel, Shelley Sacks of the U.K., and curators Heike Strelow of Germany and Tricia Watts of Los Angeles. All of their work is available online, or through the Greenmusem.org.

(6) These formal standards were the fundamental precepts of Plato and Aristotle's aesthetic of beauty.[Beardsley, M., G., 1966, Thompson 1999]

(7) Mitigation policies for wetlands trading in particular have been disastrous. There is little or no knowledge and oversight at the state and local levels when natural systems are removed and replaced to enable development

(8) I am using the standard dictionary definition of philosophical subject: that which thinks, feels, perceives, intends etc., as contrasted with the object of thought, feeling etc.

(9) I refer to general systems theory that helps us see the complexity of a problem as an interacting collection of parts which function as a singular whole.

(10) Allen Carlson's models for the aesthetic appreciation of nature. (2000: 6-8)
1, the Formal object/landscape models – The appreciation of identifiable objects within a landscape or as a scene carefully framed and chosen for consideration.
2, Metaphysical imagination model – Aesthetic appreciation as deep meditation and wild speculation. An attempt to learn the "true character of nature and our proper place in its grand design."
The first model makes the case for an environmental aesthetic by neglecting normal experience and the second raises the question of the nature-culture relationship. Carlson considers neither to be plausible contemporary models.
3, Natural environmental model –"The appreciation of nature for what it is, and for what we can know about it through the natural sciences – accommodating its true character as well as our normal experience and understanding of it."
4, Arousal model – Appreciation of nature through emotional arousal. "This less intellectual more visceral experience of nature is a way to legitimately appreciate nature without involving any knowledge gained from science."
5, Pluralist model – Acceptance of the post-modern range of ideas that attend nature, qualifying them with "serious appropriate interpretation".

This next grouping provides the working set for his decisions. The third, the intellectual is qualified by the fourth the emotional, and the fifth, a (modified post-modern) pluralist model, provides permission for qualified consideration of both approaches to knowledge.
6, Engagement model – Absorbs the appreciator into the natural environment. This model is intended to remove the traditional dichotomies of subject and object.
7, Mystery model –The only appropriate aesthetic experience of nature is based on its mystery, an appreciative incomprehension which can only come from separation from nature.
8, Non-aesthetic model –Based on the view that aesthetic appreciation is directly tied to human artifacts, therefore the aesthetic appreciation of nature is impossible.
9, Post-modern model – Art, experience, knowledge, literature, myth, science and stories all inform our aesthetic appreciation of nature, with none weighted above or below the other.
The final grouping are considered out of the question for Carlson. These are models which help define his understanding of the limits of aesthetic appreciation. Briefly, the sixth identifies a need to retain the subject-object dichotomy, because its loss negates aesthetic interpretation, the seventh clarifies the point that one cannot appreciate what one does not know, the eighth is merely self-canceling, and finally, he deems the ninth unworkable due to open-ended multiple source interpretation without qualifiers.

(11) In 1990 a three day meeting was held at the Aspen Institute. This definition was accepted by the workshop participants. It defines health in terms of four major characteristics relevant to complex systems, sustainability, activity, organization and resilience.

(12) I came across this term as a natural part of the inquiry, thinking and wordplay that occurs while writing an article like this. Realizing its obvious logical application I did a little research and discovered that ecohumanism is a concept that has emerged in the humanist literature. A very brief overview of that literature indicates that it has a more anthropocentric and theologic intention, in that context. Without getting into a detailed analysis of the term, its definitions, and its emergent literature, what seems to be consistent is the sense of limits and responsibility which emerge from an integration of nature and culture. I intend to think more about this in the future, considering the position of fusion rather than integration and what that means in terms of the artist's work on issues of individual perception, values and impact on society and politics. The text that I found most relevant to the concept, is Robert B. Tapp, *Ecohumanism:* Vol 15, of *Humanism Today*, Prometheus Books, N.Y..

Adorno, T., W. *Aesthetic Theory*. trans. Hullot-Kentor, R., University of Minnesota Press, Minneapolis, MN., (1997) 231p

Auping, M., *Earth Art: A Study in Ecological Politics* in ed. Sonfist, A., (1983) *Art in the Land*, E.P. Dutton, Inc. New York, N.Y., (1983) P. 99

Beardsley, J., *Probing the Earth: Contemporary Land Projects*, Hirschorn Museum, Smithsonian Institution, Washington D.C.(1977)

Beardsley, J., *Earthworks and Beyond*, Abbeville Press, New York, N.Y., (1984)

Beardsley, M., G., *Aesthetics from Classic Greece to the Present, Macmillan* Co. New York, New York. (1966) P43 and 61.

Berleant, A., *The Aesthetics of Environment*, Temple University Press, Philadelphia, PA., (1992)

Bookchin, M., *Our Synthetic Environment*, Colophon, N.Y., N.Y., (1974) P. xv

Bookchin, M., *Toward an Ecological Society*, Black Rose Books, Montreal, Canada (1980)

Bookchin, M., *The Ecology of Freedom*, Cheshire Books, Palo Alto CA., (1982)

Bradshaw, A.D., *Goals of Restoration,* in Ed. Gunn, J.M., *Restoration and Recovery of an Industrial Region,* Springer-Verlag New York, N.Y., (1995) P. 105

Burnham, J., *Great Western Salt Works: Essays on the Meaning of Post-Formalist Art,* George Braziller, Inc., N.Y. NY., (1974) P. 15-24.

Carlson, A., *Aesthetics and the Environment: The Appreciation of Nature, Art and Architecture, Routledge,* London, (2000)

Costanza, R., *Towards and Operational Definition of Ecosystem Heatlh,* in, Eds, Costanza, R., Norton, B.G., Haskell, B.D. *Ecosystem Health: New Goals for Environmental Management.,* Island Press, Washington D.C., (1992)

Danto, A.,C., *After the End of Art: Contemporary Art and the Pale of History,* the A.W. Mellon Lectures in the Fine Arts, 1995, Bollingen Series XXXV: 44, Princeton University Press, New Jersey, (1997)

Eaton, M., M., *The Beauty that Requires Health, in Placing Nature: Culture and Landscape Ecology.* Ed. Nassauer, J.I.,Washington D.C. Island Press, (1997) P. 85-105

Elliot, R., *Faking Nature,* (originally published in 1982) in Ed. Throop, W., *Environmental Restoration: Ethics, Theory and Practice,* Humanity Books, an imprint of Prometheus Books, Amherst, N.Y. (2000)

Foster, C., *Restoring Natue in American Culture: An Environmental Aesthetic Perspective,* in eds., Gobster, P.H., Hull, R.B., *Restoring Nature,* Island Press, Washington. D.C., (2000) P. 71-94

Ga blik, S., *Has Modernism Failed?,* Thames and Hudson, New York, N.Y., (1984)

Hardin, G., *The Tragedy of the Commons.* Science, 162 (1968), P. 1243-1248

Haskel, B.D., Norton, B.G., Costanza, R., *What is Ecosystem Health and Why Should We Worry About It?* Island Press, Washington D.C. (1992) P. 9

Hays, S., *Conservation and the Gospel of Efficiency,* Harvard University Press, Boston Mass. (1959) P. 123

Hobbs, R., *Robert Smithson: Sculpture,* Cornell University Press, London, U.K., (1981)

Jordan, W. R. III, *A Perspective of the Arboretum at 50, in Our First Fifty Years: The University of Wisconsin, Madison Arboretum. 1934-1984.* http:///libtext.library.wisc.edu/Arboretum/, Currently only accessible through http://wiscino.doit.wisc.edu/arboretum then click on the link to Historic Arboretum Documents. ©Board of Regents of the University of Wisconsin. (1984) P. 23-24

Eds., Jordan, W.R. III, Gilpin, M.E., Aber, J.D., *Restoration Ecology: A Synthetic Approach to Ecological Research,* Cambridge University Press, Cambridge, U.K., (1987)

Katz, E., *The Big Lie: Human Restoration of Nature,* in Ed. Throop, W., *Environmental Restoration: Ethics, Theory and Practice,* Humanity Books, an imprint of Prometheus Books, Amherst, N.Y. (2000) P. 83-94

Ed. Kastner, J., Wallis, B., *Land and Environmental Art,* Phaidon Press, London, U.K., (1998)

Kosuth, J., *Art After Philosophy and After: Collected Writings,* 1966-1990, MIT Press, Cambridge Mass., (1993)

Light, A., *Restoration or Domination: A Reply to Katz,* in Ed. Throop, W., Environmental Restoration: Ethics, Theory and Practice, Humanity Books, an imprint of Prometheus Books, Amherst, N.Y. (2000) P. 95-111

Lippard, L. R., *Overlay: Contemporary Art and the Art of PreHistory,* Pantheon Books, New York, N.Y., (1983)

Margulis, L., Sagan, D., *Microcosmos: Four Billion Years of Evolution from our Microbial Ancestors,* University of California Press, Berkeley and Los Angeles, CA, (1997) P. 56

Marturana, H.R., and F.J. Varela. 1980. *Autopoiesis: The Organization of the Living.* In Marturana, H.R., and F.J. Varela, Autopoiesis and Cognition. Boston: D. Reidel..

Marturana H. R., Varela F. J., *The Tree of Knowledge: The Biological Roots of Human Understanding,* Shambala Publications, Inc. Boston, Mass. (1987) P. 245

Matilsky, B., *Fragile Ecologies: Contemporary Artists Interpretations and Solutions,* Rizzoli International Publications, New York, N.Y., (1992)

Merchant, C., *The Death of Nature: Women, Ecology and the Scientific Revolution,* Harper and Row, New York, (1983)

Merchant, C., *Radical Ecology: The Search for a Livable World,* Routledge, London and New York, (1992)

Merchant, C., *Earthcare: Women and the Environment,* Routledge, London and New York, (1996)

Mitchell, J.H. *The Wildest Place on Earth: Italian Gardens and the Invention of Wilderness,* Counterpoint Press, Perseus Books, Philadelphia, PA , (2001)

Morris, R., *Robert Morris Keynote Address, in Earthworks: Land Reclamation as Sculpture,* Seattle Art Museum, Seattle WA., (1979) P. 11-16

Morris, R., *Notes on Art as/and Land Reclamation,* in *Continuous Project Altered Daily: The Writings of Robert Morris.* Massachusetts Institute of Technology, Boston Mass., (1993) P. 211-232

Naess, A., *Ecology Community and Lifestyle,* Ed/Trans Rothenberg, D., Cambridge University Press, N.Y., N.Y., (1989) P. 29E

Nassauer, J.I., *Introduction: Culture and Landscape Ecology: Insights for Action,* In *Placing Nature*: Culture and Landscape Ecology. Ed. Nassauer, J.I., Washington D.C. Island Press. (1997) P. 3-11

Nassauer, J.I., *Cultural Sustainability: Aligning Aesthetics and Ecology,* In *Placing Nature: Culture and Landscape Ecology.* Ed. Nassauer, J.I., Washington D.C. Island Press. (1997) P. 65-83

Oakes, B., *Sculpting with the Environment,* Van Nostrand Reinhold, New York, N.Y., (1995)

Plumwood, V., *Feminism and the Mastery of Nature,* Routledge, London, U.K., (1993) P. 198

Sessions, G., *Deep Ecology for the 21st Century,* Shambala Publications, Boston Mass., (1995)

Sonfist, A., *Art in the Land : A Critical Anthology of Environmental Art,* E.P. Dutton Inc., New York, N.Y., (1983)

Spaid, S., Lipton, A., *Ecovention: Current Art to Transform Ecologies,* Co-publisihed by the Cincinnati Art Center, Ecoartspace and the Greenmuseum.org., (2002)

Strelow, H., *Natural Reality: Artistic Positions Between Nature and Culture.* Ludwig Forum for Internationale Kunst, Daco Verlag, Stuttgart. (1999)

Thompson, I., *Ecology Community and Delight: Sources of Value in Landscape Architecture,* Routledge, London, U.K. (1999) P. 15

Tapp, R.B., *Ecohumanism: Vol 15, of Humanism Today,* Prometheus Books, N.Y.. (2002)

Chapter 10: Mozingo

Note: Four accompanying images are available from the author.

(1) Alan Pred proposes the "daily path" as away of understanding the iterative nature of the relationship between social structures and spatial structures, between culture and place. The daily path is the fundamental unit of time geography, what a person encounters in ordinary life, how social values shape and are shaped by the experience, and how, collectively, societies form social systems, which, according to Pred, are "place-particular". (Pred 1990)

(2) Strangely enough, the journalist Michael Pollan in his book *Second Nature: A Gardener's Education* (1991) utilizes Cicero's term without citing or discussing its origin. Its insight lies in how a non-specialized yet educated and aware citizen is grasping the import of decisions about changes in the landscape.

(3) This text on the significance of aesthetics in ecological design is based on my article "The Aesthetics of Ecological Design: Seeing Science as Culture" Landscape Journal 1997 which contains an expanded discussion of the ideas presented here.

(4) Richard Forman made this comment at the plenary session of the International Association of Landscape Ecology held at the University of Minnesota in April, 1995.

(5) For an illuminating assessment of the ecological value of the restoration of a fragment of an urban stream see Purcell, et al, 2002.

(6) According to Michael Boland, landscape architect for the Presidio Trust and project manger of the Crissy Field design, and Jeff Haltiner, senior engineer with Phillip Williams Associates and a veteran of many successful restoration projects in the Bay Area, the marsh is simply too small too establish ecological function, something the landscape architects, Hargreaves Associates, were warned about. The Presidio Trust is actively working to redress the problem.

Attfield, Robin. 2000. "Rehabilitating Nature and Making Nature Habitable." In *Environmental Restoration: Ethics, Theory, and Practice,* edited by William Throop. New York: Prometheus Books. 83-93. First published in *Philosophy and the Natural Environment,* edited by R. Attfield and A. Belsey. Cambridge: Cambridge University Press, 1994.

Baker, Elizabeth C. 1983. "Artworks on the Land." in Alan Sofist ed. *Artwork in the Land: A Critical Anthology of Environmental Art.* New York: E.P. Dutton: 73-84.

Brunson. 2000. "Managing Naturalness as a Continuum." In *Restoring Nature: Perspectives from the Social Sciences and Humanities,* edited by Paul H. Gobster and R. Bruce Hull. Covelo, California: Island Press. 229-246.

Cronon, William, editor. 1996. *Uncommon Ground: Rethinking the Human Place in Nature.* New York: W. W. Norton & Company.

Elliot, Robert. 2000. "Faking Nature." In *Environmental Restoration: Ethics, Theory, and Practice,* edited by William Throop. New York: Prometheus Books. 71-82. First published in Inquiry 25 (1982):81-93.

Foster, Cheryl. 2000. "Restoring Nature in American Culture: An Environmental Aesthetic Perspective." In *Restoring Nature: Perspectives from the Social Sciences and Humanities,* edited by Paul H. Gobster and R. Bruce Hull. Covelo, California: Island Press. 71-96.

Glacken, Clarence. 1967. *Traces on the Rhodian Shore: Nature and Culture in Western Thought from Ancient Times to the End of the Eighteenth Century.* Berkeley: University of California Press.

Gobster, Paul. 2000. "Restoring Nature: Human Actions, Interactions, and reactions." In *Restoring Nature: Perspectives from the Social Sciences and Humanities,* edited by Paul H. Gobster and R. Bruce Hull. Covelo, California: Island Press. 1-20.

Gobster, Paul and Susan C. Barro. 2000. "Negotiating Nature: Making Resoration Happen in an Urban Park Context." In *Restoring Nature: Perspectives from the Social Sciences and Humanities,* edited by Paul H. Gobster and R. Bruce Hull. Covelo, California: Island Press. 185-207.

Helford, Reid M. 2000. "Constructing Nature as Constructing Science: Expertise, Activist Science, and Public Conflict in the Chicago Wilderness." In *Restoring Nature: Perspectives from the Social Sciences and Humanities,* edited by Paul H. Gobster and R. Bruce Hull. Covelo, California: Island Press. 119-142.

Hull, R. Bruce and David P. Robertson. "The Language of Nature Matters: We Need More Public Ecology." In *Restoring Nature: Perspectives from the Social Sciences and Humanities,* edited by Paul H. Gobster and R. Bruce Hull. Covelo, California: Island Press. 185-207.

Hough, Michael. 1995. *Cities and Natural Process.* New York: Routledge.

Jacobs, Jane. 1961. *The death and Life of Great American Cities.* New York; random House.

Jordan, William R. 2000. "'Sunflower Forest' Ecological Restoration as the Basis for a New Environmental Paradigm." In *Environmental Restoration: Ethics, Theory, and Practice,* edited by William Throop. New York: Prometheus Books. 205-220. First published in *Beyond Preservation: Restoring and Inventing Landscapes,* edited by A. Dwight Baldwin, Judith De Luce, and Carl Pletsch. Minneapolis: University of Minnesota Press, 1994.

Kaplan, Rachel and Stephen Kaplan. 1989. *The Experience of Nature: a Psychological Perspective.* New York: Cambridge University Press.

Kaplan, Rachel, Stphen Kaplan, and Robert L. Ryan. 1998. *With People in Mind: Design and Managment of Everyday Nature.* Covelo, California: Island Press.

Katz, Eric. 2000. "Another look at Restoration;Technology and Artificial Nature." In *Restoring Nature: Perspectives from the Social Sciences and Humanities,* edited by Paul H. Gobster and R. Bruce Hull. Covelo, California: Island Press. 37-48.

Katz, Eric. 2000. "The Big Lie: Human Restoration of Nature." In *Environmental Restoration: Ethics, Theory, and Practice,* edited by William Throop. New York: Prometheus Books. 83-93. First published in *Research in Philosophy and Technology* (1992).

Light, Andrew. 2000. "Restoration or Domination: A Reply to Katz." In *Restoring Nature: Perspectives from the Social Sciences and Humanities,* edited by Paul H. Gobster and R. Bruce Hull. Covelo, California: Island Press. 95-111.

Merchant, Carolyn. 1989. *The Death of Nature.* San Francisco: Harper Collins.

Mozingo, Louise A. 1997. "The Aesthetics of Ecological Design: Seeing Science as Culture." *Landscape Journal.* 16(1): 46-59.

Nassauer, Joan Iverson. 1995. "Messy Ecosystems, Orderly Frames." *Landscape Journal.* 14(2): 161-170.

Nassauer, Joan Iverson. 1997. "Cultural Sustainability; Aligning Aesthetics and Ecology." in *Placing Nature: Culture and Landscape Ecology.* Edited by Joan Iverson Nassauer. Covelo, California: Island Press. 65-83.

Nassauer, Joan Iverson. 1992. "The Appearance of Ecological Systems as a Matter of Policy." *Landscape Ecology*. 6(4): 239-250

Orr, David. 1992. *Ecological Literacy: Education and the Transition to a Postmodern World.* Albany: State University of New York Press.

Pollan, Michael. 1991. *Second Nature: A Gardener's Education.* New York: Delta.

Pred, Alan. 1981. *Making Histories and Constructing Human Geographies: The Local Transformation of Practice, Power Relations, and Consciousness.* Boulder, Colorado: Westview Press.

Purcell, Alison, Frieidrich, Carla and Resh, Vincent. 2002. "An Assessment of a Small Urban Stream Project in Northern California." *Restoration Ecology*. 10:4. 685-694.

Rapoport, Amos. 1990. *The Meaning of the Built Environment: A Nonverbal Communication Approach.* Tucson: University of Arizona Press.

Rice, Faye. 1993. "Environmental Scorecard: The 10 Laggards." *Fortune.* 128(2):122 (July 26)

Thayer, Robert. 1989. "The Experience of Sustainable Landscapes." *Landscape Journal.* 8(2): 101-109.

Thayer, Robert. 1994. *Gray World, Green Heart: Technology, Nature and Sustainable Landscape.* New York: John Wiley.

Xueqin, Cao. 1973. *The Story of the Stone.* (also known as The Dream of the Red Chamber) Vol.1: The Golden Days. Translated by David Hawkes. New York: Penguin Books.

Chapter 11: Ryan

Adams, C. A. 2002. "Lead or Fade into Obscurity: Can Landscape Educators Ask and Answer Useful Questions about Ecology. In *Ecology and Design: Frameworks for Learning,* edited by B. R. Johnson and K. Hill. Washington, DC: Island Press.

Botkin, D, and E. Keller. 1995. *Environmental Science: Earth as a Living Planet.* New York: John Wiley & Sons.

Brown, B., T. Harkness, and D. Johnston. 1998. "Eco-Revelatory Design: Nature Constructed/ Nature Revealed. Guest Editors Introduction." *Landscape Journal* (Special issue): x-xvi.

Cronbach, L.J. 1951. Coefficient alpha and the internal structure of tests. *Psychometrika,* 16: 297-335.

Gobster, P. H., and R. B. Hull, eds. 2000. *Restoring Nature: Perspectives from the Social Sciences and Humanities.* Washington, DC: Island Press.

Jordan, III, W. R., M. E. Gilpin, and J. D. Aber. 1987. "Restoring Ecology: Ecological Restoration as a Technique for Basic Research." In *Restoration Ecology: A Synthetic Approach to Ecological Research,* edited by W. R. Jordan III, M. E. Gilpin, and J. D. Aber. New York: Cambridge University Press.

Jordan, III, W. R. 1989. "Restoring the Restorationist." *Restoration and Management Notes* 7(2):55.

Johnson, B. R. and K. Hill., eds. 2002. *Ecology and Design: Frameworks for Learning.* Washington, DC: Island Press.

Karr, J. R. 2002. "What from Ecology is Relevant to Design and Planning?" In *Ecology and Design: Frameworks for Learning,* edited by B. R. Johnson and K. Hill. Washington, DC: Island Press.

Lodwick, D. G. 1994. "Changing Worldviews and Landscape Restoration." In *Beyond Preservation: Restoring and Inventing Landscapes,* edited by A. D. Baldwin, Jr., J. De Luce, and C. Pletsch. Minneapolis: University of Minnesota Press.

Nassauer, J. I. 2002. "Ecological Science and Landscape Design: A Necessary Relationship in Changing Landscape." In *Ecology and Design: Frameworks for Learning,* edited by B. R. Johnson and K. Hill. Washington, DC: Island Press.

Nassauer, J. I. 1995. "Messy Ecosystems, Orderly Frames." *Landscape Journal* 14 (2): 161-170.

Norusis, M. J. 1993. *SPSS Professional Statistics 6.1.* Chicago, IL: SPSS Inc.

Nunnally, J.C. 1978. *Psychometric Theory.* New York: McGraw-Hill.

O'Connell, K. A. 2001. "The Art in Decomposition." *Landscape Architecture* 91 (8): 84-85, 95.

Pennsylvania Department of Environmental Protection. 2000. Pennsylvania's Land Recycling Program: Annual Report 2000. Harrisburg, PA: Pennsylvania Department of Environmental Protection.

Rolston, III, H. 1994. Conserving Natural Value. New York: Columbia University Press. Reprinted in *Environmental Restoration: Ethics, Theory, and Practice,* edited by W. Throop. New York: Humanity Books.

Ryan, R. L. 2000. "A People-Centered Approach to Designing and Managing Restoration Projects: Insights from Understanding Attachment to Urban Natural Areas." In *Restoring Nature: Perspectives from the Social Sciences and Humanities,* edited by P. H. Gobster and R. B. Hull. Washington, DC: Island Press.

Ryan, R. L., R. Kaplan, and R. E. Grese. 2001. "Predicting Volunteer Commitment in Environmental Stewardship Programmes." *Journal of Environmental Planning and Management* 44(5): 629-648.

Steingraber, S. and K. Hill. 2002. "Human Health and Design: An Essay in Two Parts." In *Ecology and Design: Frameworks for Learning,* edited by B. R. Johnson and K. Hill. Washington, DC: Island Press.

Throop, W. 1997. "The Rationale for Environmental Restoration." In *The Ecological Community,* edited by R. S. Gottlieb. New York: Routledge.

Throop, W. 2000. "Introduction: The Practice of Environmental Restoration." In *Environmental Restoration: Ethics, Theory, and Practice,* edited by W. Throop. New York: Humanity Books.

United States Environmental Protection Agency. 1997. "The Clean Water Act: A Snapshot of Progress in Protecting America's Waters." Environmental Protection Agency's Office of Water. web-site: http://www.epa.gov/owow/cwa/25report.html. (accessed on April 4, 2002)

Vining, J., E. Tyler, and B. Kweon. 2000. "Public Values, Opinions, and Emotions in Restoration Controversies." In *Restoring Nature: Perspectives from the Social Sciences and Humanities,* edited by P. H. Gobster and R. B. Hull. Washington, DC: Island Press.

Acknowledgement
A debt of gratitude goes to my long-time research assistant, Amanda Walker for her help distributing this survey, encoding the data, and running the data analyses. Thanks also goes to the other graduate students from the Department of Landscape Architecture and Regional Planning at the University of Massachusetts, Amherst who helped distribute and collect surveys for this study: Patricia Gambarini, Mark Lewis, and Mary Scipioni. Finally, thanks to Robert France at Harvard Graduate School of Design for inviting me to participate in this stimulating and creative venture; it is an honor to be able to discuss environmental restoration issues with so many talented leaders in the field.

Conclusion: France

McDonough, W. and M. Brangart. 2002. *Cradle to cradle: Remaking the way we make things.* North Point Press.

Dansereau, P. 1975. *Inscape and landscape.* Univ. Press California.

Dreisietl, H. 2003. Waterworks. Foreword *in* France, R.L. *Deep immersion: The experience of water.* Green Frigate Books.

France, R. L. 2003. *Deep immersion: The experience of water.* Green Frigate Books.

France, R.L. 2006. *Introduction to watershed development: Understanding and measuring the impacts of sprawl.* Rowman & Littlefield.

Graves, R. 1958. *Good-bye to all that: An autobiography.* Anchor.

Halifax, J. 1994. *The fruitful darkness: Reconnecting with the body of the Earth.* HarperCollins.

Higgs, E. 2003. *Nature by design: People, natural process, and ecological restoration.* MIT Press.

Hubbard, B. M. 1998. *Conscious evolution: Awakening the power of our social potential.* New World Library.

Jordan, W.R. 2003. *The sunflower forest: Ecological restoration and the new communion with nature.* Univ. Caif. Press.

Krinke, R. 2001. Overview: Design practice and manufactured sites. Pg. 125-149 *in* Kirkwood, N. (Ed.) *Manufactured sites: Rethinking the post-industrial landscape.* Spon Press.

Leopold, A. 1949. *A Sand County almanac with essays on conservation from Round River.* Ballatine Books.

McKibben, B. 1990. *The end of nature.* Anchor.

Muir, R. 1999. *Approaches to landscape.* Routledge.

Palmer, J.A. 1998. *To know as we are known: Education as a spiritual journey*. Harper.

Proust, M. 1982. *Remembrance of things past: Vol. 1-3*. Vintage.

Schama, S. 1996 . *Landscape and memory*. Vintage.

Sprin, A. 1996. Constructing nature: The legacy of Frederick Law Olmsted. Pg. 91-113 *in* Cronon, W. (Ed.) *Uncommon ground: Rethinking the human place in nature*. W.W. Norton.

Epilogue: France

Connel, E.S. 1998. *Deus lo volt! Chronicle of the crusades*. Counter Point.

France, R.L. (ED.) 2003. *Profitably soaked: Thoreau's experience of water*. Green Frigate Books.

France, R.L. (Ed.) 2007. *Wetlands of mass destruction: Ancient presage for contemporary ecocide in southern Iraq*. Green Frigate Books.

France, R.L. 2008. *Sustainable development of the Iraqi marshlands: Lessons and relevant applications from around the world*. Routledge.

France, R.L. 2009. *Restoring the Iraqi marshlands: Potentials, perspectives, and practices*. Sussex Academic Press.

France, R.L. 2009. *Back to the Garden: Searching for Eden in the Mesopotamian marshes*. Harvard University Press.

Giblet, R. 1996. *Post-modern wetlands: Culture, history, ecology*. Edinburgh Univ. Press.

Nicholson, E. and P. Clark. 2003. *The Iraqi marshlands: A human and environmental study*. Politico's Publ.

Pollan, M. 1991. *Second nature: A gardener's education*. Delta.

Schanberg, S.H. 2003. The widening crusade: Bush's war plan is scarier than he's saying. *The Village Voice* 42: 32-39.

Thesiger, W. 1964. *The marsh Arabs*. Penguin.

United Nations Environmental Protection Program. 2001. *The Mesopotamian marshlands: Demise of an ecosystem*. Internal document.

United States Agency for International Development. 2003. *Strategies for assisting the marsh dwellers and restoring the marshlands in southern Iraq*. Internal document.

Other Green Frigate Books

A Wanderer All My Days: John Muir in New England
J. Parker Huber

Ultreia! Onward! Progress of the Pilgrim
Robert Lawrence France

Uncommon Cents: Thoreau and the Nature of Business (*In Press*)
Robert M. Abbott

Wetlands of Mass Destruction: Ancient Presage for Contemporary Ecocide in Southern Iraq
Robert Lawrence France

Where We Live: Chasing the Dream of Urban Sustainability (*In Press*)
Mark Holland and Robert Abbott

Profitably Soaked: Thoreau's Experience of Water
Robert Lawrence France

Destination Mutable (*In Press*)
John Ballie

Deep Immersion: The Experience of Water
Robert Lawrence France

Liquid Gold: The Lore and Logic of Using Urine to Grow Plants (First Edition)
Carol Steinfeld

GREEN FRIGATE BOOKS

"THERE IS NO FRIGATE LIKE A BOOK"

Words on the page have the power to transport us, and in the process, transform us. Such journeys can be far reaching, traversing the landscapes of the external world and that within, as well as the timescapes of the past, present and future.

Green Frigate Books is a small publishing house offering a vehicle—a ship—for those seeking to conceptually sail and explore the horizons of the natural and built environments, and the relations of humans within them. Our goal is to reach an educated lay readership by producing works that fall in the cracks between those offered by traditional academic and popular presses.